Lecture Notes in Statistics

Edited by J. Berger, S. Fienberg, J. Gani,
K. Krickeberg, I. Olkin, and B. Singer

63

Johann Pfanzagl

Learning Resources
Brevard Community College
Cocoa, Florida

Estimation in Semiparametric Models

Some Recent Developments

Springer-Verlag
New York Berlin Heidelberg London Paris Tokyo Hong Kong

Author

Johann Pfanzagl
Mathematisches Institut der Universität zu Köln
Weyertal 86, 5000 Köln 41, Federal Republic of Germany

Mathematical Subject Classification: 62-02, 62G05

ISBN 0-387-97238-2 Springer-Verlag New York Berlin Heidelberg
ISBN 3-540-97238-2 Springer-Verlag Berlin Heidelberg New York

This work is subject to copyright. All rights are reserved, whether the whole or part of the material is concerned, specifically the rights of translation, reprinting, re-use of illustrations, recitation, broadcasting, reproduction on microfilms or in other ways, and storage in data banks. Duplication of this publication or parts thereof is only permitted under the provisions of the German Copyright Law of September 9, 1965, in its version of June 24, 1985, and a copyright fee must always be paid. Violations fall under the prosecution act of the German Copyright Law.

© Springer-Verlag Berlin Heidelberg 1990
Printed in Germany

Printing and binding: Druckhaus Beltz, Hemsbach/Bergstr.
2847/3140-543210 – Printed on acid-free paper

Contents

Introduction 1

A   Survey of basic theory
     1. Tangent spaces and gradients 2
     2. Asymptotic bounds for the concentration of
         estimator–sequences 4
     3. Constructing estimator–sequences 7
     4. Estimation in semiparametric models 17
     5. Families of gradients 23
     6. Estimating equations 35

B   Semiparametric families admitting a sufficient statistic
     7. A special semiparametric model 38
     8. Mixture models 48
     9. Examples of mixture models 53
         Example 1 67
         Example 2 75
         Example 3 82

L   Auxiliary results 88

References 106
Notation index 110
Subject index 112

Introduction

Assume one has to estimate the mean $\int x P(dx)$ (or the median of P, or any other functional $\kappa(P)$) on the basis of i.i.d. observations from P. If nothing is known about P, then the sample mean is certainly the best estimator one can think of. If P is known to be the member of a certain parametric family, say $\{P_\vartheta : \vartheta \in \Theta\}$, one can usually do better by estimating ϑ first, say by $\vartheta^{(n)}(\underline{x})$, and using $\int x P_{\vartheta^{(n)}(\underline{x})}(dx)$ as an estimate for $\int x P_\vartheta(dx)$. There is an "intermediate" range, where we know something about the unknown probability measure P, but less than parametric theory takes for granted.

Practical problems have always led statisticians to invent estimators for such intermediate models, but it usually remained open whether these estimators are nearly optimal or not. There was one exception: The case of "adaptivity", where a "nonparametric" estimate exists which is asymptotically optimal for *any* parametric submodel. The standard (and for a long time only) example of such a fortunate situation was the estimation of the center of symmetry for a distribution of unknown shape.

Starting with Levit (1974) the concepts evolved which are needed for a general (i.e. not necessarily parametric) asymptotic theory, and which led to asymptotic bounds for the concentration of estimator–sequences in general nonparametric models. The basic ideas underlying this approach are sketched in Sections 1 and 2. Section 3 indicates conditions under which asymptotically maximally concentrated estimator–sequences exist. Since the applicability of such general conditions seems to be questionable, Section 4 turns to more special "semiparametric" models where a finite dimensional parameter is to be estimated in the presence of a general "nuisance" parameter. Even in this special situation a general procedure for the construction of asymptotically optimal estimator–sequences is not in sight. Therefore Section 5 is devoted to an even more special case. The most useful applications are to families of distributions admitting a sufficient statistic for the nuisance parameter (see Section 7). As a particular case, mixture models are studied in more detail in Sections 8 and 9. The relationship to the theory of estimating equations is discussed in Section 6.

1. Tangent spaces and gradients

Let $(X, \mathcal{A})$ be a measurable space, and $\mathcal{P}$ a family of probability measures (for short: p–measures) $P|\mathcal{A}$. We consider the problem of estimating the value of a functional $\kappa : \mathcal{P} \to \mathbb{R}$, based on an i.i.d. sample $(x_1, \ldots, x_n)$, i.e. a realization from P^n, for some $P \in \mathcal{P}$. The restriction to 1–dimensional functionals makes the following presentation more transparent. It is justified by the fact that the problem of estimating an m–dimensional functional simply is the problem of estimating its m (1–dimensional) components. (The essential point: componentwise as. efficiency implies joint as. efficiency. See I *), p. 159, Corollary 9.3.6.)

To study the *asymptotic* performance of estimator–sequences (for short: e.s.) we have to seize on the *local* structure of $\mathcal{P}$, and on the *local* performance of κ.

This can be done by considering for $P \in \mathcal{P}$ paths $P_t \in \mathcal{P}$, converging to P for $t \to 0$ (written as $P_t \to P$ in the following). To keep our presentation transparent, we assume that the p–measures in $\mathcal{P}$ are mutually absolutely continuous. The density of P with respect to some dominating measure μ will be denoted by p. Given a p–measure $P \in \mathcal{P}$, let $\mathcal{L}_*(P) = \{g \in \mathcal{L}_2(P) : P(g) = 0\}$. A path $P_t \to P$ for $t \to 0$ is *differentiable* with *derivative* $g \in \mathcal{L}_*(P)$ if the densities can be represented as

$$\frac{p_t(x)}{p(x)} = 1 + tg(x) + tr_t(x) \tag{1.1}$$

with remainder term r_t converging to 0 in the following sense (see II, Section 1.2)

$$P\big(|r_t|1_{\{|r_t|>t^{-1}\}}\big) = o(t),$$
$$P\big(r_t^2 1_{\{|r_t|\leq t^{-1}\}}\big) = o(t^0).$$

As a convenient façon de parler we shall say that the path P_t, $t \to 0$, converges to P from *direction* g.

Technically speaking, the conditions on r_t, $t \to 0$, are to ensure that the "subsequence" $P_{n^{-1/2}}$, $n \in \mathbb{N}$, selected from this path fulfills an LAN–condition,

---

*) see Notation index

i.e.
$$\sum_1^n \log[p_{n^{-1/2}}(x_\nu)/p(x_\nu)]$$
$$= n^{-1/2} \sum_1^n g(x_\nu) - \frac{1}{2} P(g^2) + o_p(n^0) \qquad (P^n). \tag{1.2}$$

The *tangent space* $T(P,\mathcal{P})$ of $\mathcal{P}$ at P is the set of all functions $g \in \mathcal{L}_*(P)$ occurring as derivatives of paths in $\mathcal{P}$ converging to P. Throughout the following we assume that $T(P,\mathcal{P})$ is a closed linear subspace of $\mathcal{L}_*(P)$, which holds true in all natural examples.

The functional $\kappa : \mathcal{P} \to \mathbb{R}$ is *differentiable* at P if there exists a function $\kappa^+(\cdot, P) \in \mathcal{L}_*(P)$ with the following property: For every $g \in T(P,\mathcal{P})$ there exists a path $P_t \to P$ with derivative g such that

$$\kappa(P_t) - \kappa(P) = tP\big(\kappa^+(\cdot, P)g\big) + o(t). \tag{1.3}$$

Any function $\kappa^+(\cdot, P)$ with this property is called a *gradient* of κ at P. Obviously, a gradient is not unique. Since $\kappa^+(\cdot, P)$ enters (1.3) through $P\big(\kappa^+(\cdot, P)g\big)$ only, any function differing from $\kappa^+(\cdot, P)$ by a function in $\mathcal{L}_*(P)$ orthogonal to $T(P,\mathcal{P})$ serves the same purpose. There is, however, a distinguished gradient, the one in $T(P,\mathcal{P})$. It will be called the *canonical gradient*, and denoted by $\kappa^*(\cdot, P)$. Since $T(P,\mathcal{P})$ is a closed linear space, such a canonical gradient always exists. It can be obtained as the projection of any gradient $\kappa^+(\cdot, P)$ into $T(P,\mathcal{P})$, so that

$$P\big(\kappa^*(\cdot, P)^2\big) \leq P\big(\kappa^+(\cdot, P)^2\big) \quad \text{for any gradient } \kappa^+(\cdot, P). \tag{1.4}$$

The concepts introduced here go back to papers by Levit (1974, 1975), Koshevnik and Levit (1976). They have been further developed in I and II, Begun et al. (1983) and van der Vaart (1988). See also the forthcoming book by Bickel et al. (199?).

2. Asymptotic bounds for the concentration of estimator–sequences

Equipped with the concepts of "tangent space" and "gradient" we now turn to the problem of estimating the functional κ, based on an i.i.d. sample $x_1, \ldots, x_n$ generated by some $P \in \mathcal{P}$.

An *estimator* is a measurable map $\kappa^{(n)} : X^n \to \mathbb{R}$.

The relevant characteristic of an estimator $\kappa^{(n)}$ is its distribution under P^n. Except for some special cases, this distribution is not known explicitly, and different estimators cannot be evaluated with respect to their concentration about $\kappa(P)$. Therefore, the theory turns to estimator–sequences, the distribution of which can be approximated by some limiting distribution. In technical terms,

$$P^n * n^{1/2}\big(\kappa^{(n)} - \kappa(P)\big) \Rightarrow M_P \qquad (2.1)$$

(with "$\Rightarrow$" denoting weak convergence).

An estimator–sequence is *regular* if for any direction $g \in T(P, \mathcal{P})$ there is a path $P_t \to P$ such that (2.1) holds with P replaced on the left–hand side by $P_{n^{-1/2}}$ (and with M_P on the right–hand side remaining unchanged). Most e.s. occurring in literature have this property, but little can be said for defending it as a general requirement for *all* e.s.

By the Convolution Theorem (see e.g. I, p. 158, Theorem 9.3.1 for an appropriate version) the limiting distribution M_P of regular e.s. cannot be more concentrated on intervals about 0 than the normal distribution with mean 0 and variance $P\big(\kappa^*(\cdot, P)^2\big)$ (called "as. variance bound" in the following). This demonstrates the role of the canonical gradient in estimation theory.

An e.s. $\kappa^{(n)}$, $n \in \mathbb{N}$, is *asymptotically linear* with influence function $K(\cdot, P) \in \mathcal{L}_*(P)$ if it admits a representation of the following kind

$$n^{1/2}\big(\kappa^{(n)}(\underline{x}) - \kappa(P)\big) = n^{-1/2} \sum_{1}^{n} K(x_\nu, P) + o_p(n^0) \quad (P^n). \qquad (2.2)$$

Most e.s. studied in literature are of this type. The limiting distribution is the normal distribution with mean 0 and variance $P\big(K(\cdot, P)^2\big)$.

As. linear e.s. of differentiable functionals are regular iff the influence function is a gradient. This seems to be generally known. We briefly indicate the proof, since none is available in literature.

The implication "regular ⇒ gradient" occurs in Bickel (1981, p. 17, Lemma 1) for the parametric case, and for the general case in I, p. 209, Proposition 11.6.2. (Observe there is a misprint. Read "e.s. $\kappa^{(n)}$" instead of "functional κ".) The implication "gradient ⇒ regular" occurs in Rieder (1983, p. 79, relation (7)) for the parametric case.

Proposition 2.3. Let $\kappa^{(n)}$, $n \in \mathbb{N}$, be an as. linear e.s.

(i) If κ is differentiable and $K(\cdot, P)$ a gradient, then $\kappa^{(n)}$, $n \in \mathbb{N}$, is regular.

(ii) If $\kappa^{(n)}$, $n \in \mathbb{N}$, is regular, then κ is differentiable (along paths $P_{n-1/2}$) and $K(\cdot, P)$ is a gradient.

The proof is formulated for $\kappa : \mathcal{P} \to \mathbb{R}$. The result extends immediately to the case $\kappa : \mathcal{P} \to \mathbb{R}^q$ by applying it to the 1-dimensional functional $\kappa(P) = \sum_{i=1}^{q} a_i \kappa_i(P)$, with arbitrary coefficients $a_i \in \mathbb{R}$, $i = 1, \ldots, q$.

Proof. Let $P_{n-1/2}$, $n \in \mathbb{N}$, be a path with derivative $g \in T(P, \mathcal{P})$. By (2.2), the sequence of joint distributions

$$P^n * \left(n^{-1/2} \sum_{1}^{n} g(x_\nu), n^{1/2}\left(\kappa^{(n)}(\underline{x}) - \kappa(P)\right)\right), \quad n \in \mathbb{N},$$

converges weakly to a 2-dimensional normal distribution with mean 0 and covariance matrix Σ, with $\Sigma_{11} = P(g^2)$, $\Sigma_{12} = P(K(\cdot, P)g)$, $\Sigma_{22} = P(K(\cdot, P)^2)$. By (1.2) and LeCam's third lemma (see Prakasa Rao (1987), p. 104, Corollary 1.11.11 for a convenient version) this implies

$$P^n_{n-1/2} * n^{1/2}\left(\kappa^{(n)} - \kappa(P)\right) \Rightarrow N_{(\Sigma_{12}, \Sigma_{22})}. \tag{2.4}$$

If κ is differentiable (see (1.3)) and $K(\cdot, P)$ a gradient, we have

$$\kappa(P_{n-1/2}) = \kappa(P) + n^{-1/2} P(K(\cdot, P)g) + o(n^{-1/2}).$$

Hence

$$P^n_{n-1/2} * n^{1/2}\left(\kappa^{(n)} - \kappa(P_{n-1/2})\right) \Rightarrow N_{(\mu, \Sigma_{22})},$$

with

$$\mu = \Sigma_{12} - P(K(\cdot, P)g) = 0.$$

Hence $\kappa^{(n)}$, $n \in \mathbb{N}$, is regular.

Conversely, if $\kappa^{(n)}$, $n \in \mathbb{N}$, is regular, then (2.4) implies that

$$n^{1/2}\left(\kappa(P_{n-1/2}) - \kappa(P)\right) \to \Sigma_{12} = P(K(\cdot, P)g).$$

Since this holds for any $g \in T(P,\mathcal{P})$, κ is differentiable with gradient $K(\cdot,P)$. □

Since (see (1.4)) $P(K(\cdot,P)^2) \geq P(\kappa^*(\cdot,P)^2)$ for any gradient $K(\cdot,P)$, an as. linear e.s. with the influence function $\kappa^*(\cdot,P)$ is as. optimal in the class of all regular e.s. The Convolution Theorem implies in particular that *all* as. optimal regular e.s. are of this type.

3. Constructing estimator–sequences

The purpose of Section 2 was to justify $P(\kappa^*(\cdot,P)^2)$ as the "as. variance bound" for e.s. Often e.s. attaining this bound can be found by ad hoc methods. In many cases, such e.s. are already available, and the role of the "local theory" is confined to establish their as. optimality. This is, in particular, the case if the tangent space is full, i.e. $T(P,\mathcal{P}) = \mathcal{L}_*(P)$. Then there exists one gradient only, hence any as. linear e.s. is necessarily as. efficient.

A general method for constructing as. efficient e.s. is not available so far. In fact, the question is open in which cases the as. variance bound is attainable (see Bickel and Ritov (1989)).

Since all as. efficient regular e.s. are as. linear with influence function $\kappa^*(\cdot,P)$, the attention naturally turns to the construction of as. linear e.s. Such constructions demand an estimator of $\kappa^*(\cdot,P)$. Since the canonical gradient may be difficult to estimate (or even unknown) in certain cases, one might be willing to settle with e.s. which are subefficient, but easier to compute. For this, there is another justification besides economy. For approximations by limiting distributions, the errors for finite sample sizes are usually unknown. These errors are likely to be larger for more complex estimators. Hence one might feel uncertain whether the limiting distribution provides suitable information about the performance for moderate sample sizes in case of a very complex construction.

This motivates investigating the construction of as. linear e.s. which have an arbitrary gradient K (not necessarily $K = \kappa^*$) as influence function. For this purpose, we introduce for $K(\cdot,P) \in \mathcal{L}_*(P)$ the function

$$k(\cdot,P) := \kappa(P) + K(\cdot,P). \tag{3.1}$$

As a preparation to the following considerations we state

Lemma 3.2. *If there exists an e.s. for $\kappa(P)$ which is as. linear with influence function $K(\cdot,P)$, then there also exists an e.s. with the same influence function which is permutation invariant.*

Proof. Let $\kappa^{(n)}$, $n \in \mathbb{N}$, denote the e.s. with influence function $K(\cdot,P)$. Denote

$$R^{(n)}(x_1,\ldots,x_n,P) := n^{1/2}\Big(\kappa^{(n)}(x_1,\ldots,x_n) - n^{-1}\sum_{1}^{n} k(x_\nu,P)\Big).$$

By assumption, $R^{(n)}(\cdot, P) = o_p(n^0)$ (P^n). By Lemma L.17, the sequence of the medians of $\{R^{(n)}(\pi_n \underline{x}, P) : \pi_n \in \Pi_n\}$ (where Π_n denotes the class of all permutations $\pi_n : X^n \to X^n$) converges stochastically to 0 under P^n. Therefore, the sequence of the medians of $\{\kappa^{(n)}(\pi_n \underline{x}) : \pi_n \in \Pi_n\}$ is as. linear with influence function $K(\cdot, P)$. Since it is permutation invariant, this proves the assertion. □

In the following, $k^{(n)} : X^{1+n} \to \mathbb{R}$ denotes a function for which $X^n \ni \underline{x} \to k^{(n)}(\cdot, \underline{x})$ is used as an estimator for $k(\cdot, P)$.

Proposition 3.3. If

$$n^{-1/2} \sum_{1}^{n} \left(k^{(n)}(x_\nu, \underline{x}) - k(x_\nu, P)\right) = o_p(n^0) \quad (P^n), \tag{3.4}$$

then

$$\kappa^{(n)}(\underline{x}) := n^{-1} \sum_{1}^{n} k^{(n)}(x_\nu, \underline{x}), \quad n \in \mathbb{N}, \tag{3.5}$$

is as. linear with influence function $K(\cdot, P)$.

Proof.

$$n^{1/2}\left(\kappa^{(n)}(\underline{x}) - \kappa(P)\right) = n^{-1/2} \sum_{1}^{n} \left(k^{(n)}(x_\nu, \underline{x}) - \kappa(P)\right)$$

$$= n^{-1/2} \sum_{1}^{n} K(x_\nu, P) + o_p(n^0) \quad (P^n).$$

□

Observe that (3.5) is nothing else but a version of the one–step improvement procedure.

To find an e.s. $k^{(n)}$ fulfilling (3.4) and to compute (3.5) is certainly not always the best way to obtain estimators $\kappa^{(n)}$. None the less it might be of theoretical interest, at least, that $k(\cdot, P)$ is necessarily estimable in the strong sense (3.4) whenever an as. linear e.s. for $\kappa(P)$ exists. In other words: If an as. linear e.s. exists, it can always be obtained by an improvement procedure (3.5).

Proposition 3.6. Assume that there exists an e.s. $\kappa^{(n)}$, $n \in \mathbb{N}$, which is as. linear with influence function $K(\cdot, P)$. Then there exists a permutation invariant e.s. $\underline{x} \to P_{\underline{x}}^{(n)} \in \mathcal{P}$ for P such that $k^{(n)}(\cdot, \underline{x}) := k(\cdot, P_{\underline{x}}^{(n)})$ fulfills (3.4).

As a particular consequence: Any as. linear e.s. is up to $o_p(n^{-1/2})$ of the type $n^{-1} \sum_{1}^{n} k(x_\nu, P_{\underline{x}}^{(n)})$.

Proof. According to Lemma 3.2 we may assume w.l.g. that $\kappa^{(n)}(x_1,\ldots,x_n)$ is permutation invariant. For $P \in \mathcal{P}$, let

$$\Delta^{(n)}(\underline{x}, P) := n^{1/2}\Big(\kappa^{(n)}(\underline{x}) - n^{-1}\sum_{1}^{n} k(x_\nu, P)\Big).$$

By assumption, $\Delta^{(n)}(\cdot, P) = o_p(n^0) \quad (P^n)$.

For $\underline{x} \in X^n$ choose $P_{\underline{x}}^{(n)} \in \mathcal{P}$ such that

$$|\Delta^{(n)}(\underline{x}, P_{\underline{x}}^{(n)})| \leq n^{-1} + \inf_{Q \in \mathcal{P}} |\Delta^{(n)}(\underline{x}, Q)|.$$

Since $\underline{x} \to \Delta^{(n)}(\underline{x}, P)$ is permutation invariant, $\underline{x} \to P_{\underline{x}}^{(n)}$ can be chosen permutation invariant, too. We obtain

$$n^{-1/2}|\sum_{1}^{n}\big(k(x_\nu, P_{\underline{x}}^{(n)}) - k(x_\nu, P)\big)|$$
$$\leq |\Delta^{(n)}(\underline{x}, P_{\underline{x}}^{(n)})| + |\Delta^{(n)}(\underline{x}, P)|$$
$$\leq n^{-1} + \inf_{Q \in \mathcal{P}} |\Delta^{(n)}(\underline{x}, Q)| + |\Delta^{(n)}(\underline{x}, P)|$$
$$\leq n^{-1} + 2|\Delta^{(n)}(\underline{x}, P)| = o_p(n^0) \quad (P^n).$$

□

Taking Propositions 3.3 and 3.6 together, one could say that the existence of an e.s. $k^{(n)}(x,\cdot)$ for $k(x,P)$ fulfilling (3.4) is *necessary* and *sufficient* for the existence of an as. linear e.s. for $\kappa(P)$ with influence function $K(\cdot,P)$. But such a euphemistic statement conceals that we are unable to solve the relevant problem, that is, to find a method for *constructing* such an e.s. $k^{(n)}(x,\cdot)$.

This task has, of course, been solved for many special cases, in particular for certain semiparametric models. The classical example is the estimation of the center of symmetry for a p–measure on $I\!B$ with unknown symmetric Lebesgue–density (see Stone (1975), Beran (1978)). For other examples see Section 9.

If we wish to prove that a particular e.s. $k^{(n)}$ fulfills (3.4) we encounter two problems. First, the functions $(x_1,\ldots,x_n) \to k^{(n)}(x_\nu,\underline{x}) - k(x_\nu,P)$, $\nu = 1,\ldots,n$, are not stochastically independent, even though the dependence will usually be weak. Second — neglecting the question of dependence for a moment — the expectation $\int\big(k^{(n)}(x_\nu,\underline{x}) - k(x_\nu,P)\big)P(dx_\nu)$ $(= \int k^{(n)}(x_\nu,\underline{x})P(dx_\nu) - \kappa(P))$ has to be stochastically small, say $o_p(n^{-1/2})$.

Remark 3.7. The existence of e.s. $k^{(n)}$ with expectation close to $\kappa(P)$ is plausible, because $P(k(\cdot,\hat{P}))$ deviates from $\kappa(P)$ only by a small amount if $\hat{P} \in \mathcal{P}$ is close to P. Therefore, $k^{(n)}(\cdot,\underline{x}) := k(\cdot, P_{\underline{x}}^{(n)})$ (with $P_{\underline{x}}^{(n)} \in \mathcal{P}$ an estimate of P) is a candidate for an e.s. with expectation as. close to $\kappa(P)$.

To justify this claim, assume that $P \to K(x,P)$ is differentiable for every $x \in X$, with gradient $y \to \hat{K}(x,y,P)$. Then $P \to k(x,P)$ is differentiable for every $x \in X$, with gradient $y \to \hat{k}(x,y,P)$ given by

$$\hat{k}(x,y,P) = K(y,P) + \hat{K}(x,y,P).$$

By II 4.1.13 (see also I 11.5.3), the function

$$y \to \int \hat{k}(x,y,P)P(dx) \quad \text{is orthogonal to } T(P,\mathcal{P}).$$

If $y \to \hat{k}(x,y,P)$ is the canonical gradient of $P \to k(x,P)$, this implies

$$\int \hat{k}(x,y,P)P(dx) = 0 \qquad \text{for all } y \in X. \tag{3.8}$$

Therefore

$$P(k(\cdot, P_t)) = \kappa(P) + o(t) \tag{3.9}$$

for any path $P_t \to P$ (with the emphasis on $o(t)$ in contrast to $O(t)$).

In applications to special problems one can use the particular structure of $k^{(n)}$ in order to prove that the correlation between $k^{(n)}(x_\nu,\underline{x}) - k(x_\nu,P)$ and $k^{(n)}(x_\mu,\underline{x}) - k(x_\mu,P)$ is of order $o(n^{-1})$ for $\nu \neq \mu$. In the absence of such a specific knowledge one may use the splitting trick to cope with the problem of dependence.

Our starting point is that relation (3.4) may be easier to prove if the random variables x_ν occurring in the summation are stochastically independent of $\underline{x}$ occurring in the estimate $k^{(n)}(\cdot,\underline{x})$.

We assume

$$\int \bigl(k^{(n)}(x,\underline{x}) - k(x,P)\bigr) P(dx) = o_p(n^{-1/2}) \quad (P^n), \tag{3.10'}$$

$$\int \bigl(k^{(n)}(x,\underline{x}) - k(x,P)\bigr)^2 P(dx) = o_p(n^0) \quad (P^n). \tag{3.10''}$$

Now we "split" the sample $(x_1,\ldots,x_n)$, use $(x_{m_n+1},\ldots,x_n)$ for estimating $k(\cdot,P)$, and extend the summation over $x_1,\ldots,x_{m_n}$. Thanks to the stochastic independence between the two samples we obtain

$$m_n^{-1/2} \sum_1^{m_n} \bigl(k^{(n-m_n)}(x_\nu; x_{m_n+1},\ldots,x_n) - k(x_\nu,P)\bigr) = o_p(n^0) \quad (P^n), \tag{3.11}$$

provided m_n/n, $n \in \mathbb{N}$, is bounded away from 0 and $n - m_n \to \infty$.

Because of (3.10), this relation follows immediately from Lemma L.12, applied with

$$\Delta_m(x; x_1, \ldots, x_m) := k^{(m)}(x; x_1, \ldots, x_m) - k(x, P)$$
$$- \int \left(k^{(m)}(\xi; x_1, \ldots, x_m) - k(\xi, P) \right) P(d\xi).$$

Proposition 3.12. If there exists an e.s. $\underline{x} \to k^{(n)}(\cdot, \underline{x})$ for $k(\cdot, P)$ fulfilling condition (3.11), then there exists an e.s. for $\kappa(P)$ which is as. linear with influence function $K(\cdot, P)$.

Proof. To write the definition of $\kappa^{(n)}$ in a transparent way we introduce

$$k_\nu^{(n)}(\underline{x}) := \begin{cases} k^{(n-m_n)}(x_\nu; x_{m_n+1}, \ldots, x_n) & \nu \in \{1, \ldots, m_n\} \\ k^{(m_n)}(x_\nu; x_1, \ldots, x_{m_n}) & \nu \in \{m_n+1, \ldots, n\}. \end{cases} \quad (3.13)$$

From (3.11) and the corresponding relation — with $(x_1, \ldots, x_{m_n})$ and $(x_{m_n+1}, \ldots, x_n)$ interchanged — we obtain

$$n^{-1/2} \sum_1^n \left(k_\nu^{(n)}(\underline{x}) - k(x_\nu, P) \right) = o_p(n^0) \quad (P^n). \quad (3.14)$$

Hence the e.s. defined by

$$\kappa^{(n)}(\underline{x}) := n^{-1} \sum_1^n k_\nu^{(n)}(\underline{x}) \quad (3.15)$$

fulfills

$$n^{1/2} \left(\kappa^{(n)}(\underline{x}) - \kappa(P) \right) = n^{-1/2} \sum_1^n K(x_\nu, P) + o_p(n^0) \quad (P^n).$$

□

This splitting trick was used in connection with semiparametric models by Schick (1986) and Klaassen (1987). If instead of (3.10') the following stronger condition

$$\int \left(k^{(n)}(x, \underline{x}) - k(x, P) \right) P(dx) = 0 \qquad \text{for } n \in \mathbb{N} \quad (3.16)$$

holds, it suffices to use a comparatively small part of the sample for estimating $k(\cdot, P)$, i.e. one may choose $m_n \in \{1, \ldots, n\}$ such that $n - m_n \to \infty$, but $n^{-1} m_n \to 1$, and define

$$\kappa^{(n)}(\underline{x}) := m_n^{-1} \sum_1^{m_n} k^{(n-m_n)}(x_\nu; x_{m_n+1}, \ldots, x_n). \tag{3.17}$$

In this version, the splitting procedure was used by Hájek (1962, p. 1144) and by many other authors since.

*   *

*

In dealing with special examples one can usually avoid the splitting trick. Schick (1987, p. 95, Lemma 3.1) provides conditions which have been successfully applied in a number of cases. The following lemma is a variant of Schick's Lemma 3.1.

Lemma 3.18. Assume that $k^{(n)}(x; x_1, \ldots, x_n)$, permutation invariant in $(x_1, \ldots, x_n)$, fulfills (3.10') and the following stronger version of (3.10'')

$$\int \left(k^{(n)}(x, \underline{x}) - k(x, P)\right)^2 P(dx) P^n(d\underline{x}) = o(n^0). \tag{3.19}$$

Assume in addition

$$\int \left(k^{(n)}(x; x_1, x_2, \ldots, x_n) - \int k^{(n)}(x; y_1, x_2, \ldots, x_n) P(dy_1)\right)^2 \tag{3.20}$$

$$P(dx) P(dx_1) \ldots P(dx_n) = o(n^{-1}).$$

Then

$$\kappa^{(n)}(\underline{x}) := n^{-1} \sum_1^n k^{(n-1)}(x_\nu; x_1, \ldots, x_{\nu-1}, x_{\nu+1}, \ldots, x_n) \tag{3.21}$$

is as. linear with influence function $K(\cdot, P)$.

Addendum. *If, in addition, Schick's Condition (3.3) holds, i.e. if*

$$n^{-1/2} \sum_1^n \left(k^{(n-1)}(x_\nu; x_1, \ldots, x_{\nu-1}, x_{\nu+1}, \ldots, x_n) - k^{(n)}(x_\nu, \underline{x})\right) \tag{3.22}$$

$$= o_p(n^0) \quad (P^n),$$

then $\kappa^{(n)}(\underline{x}) = n^{-1} \sum_1^n k^{(n)}(x_\nu, \underline{x})$ is as. linear with influence function $K(\cdot, P)$.

Proof. With $x_{n \cdot \nu} := (x_1, \ldots, x_{\nu-1}, x_{\nu+1}, \ldots, x_n)$ and

$$\Delta_m(x; x_1, \ldots, x_m) := k^{(m)}(x; x_1, \ldots, x_m) - k(x, P)$$
$$- \int \left(k^{(m)}(\xi; x_1, \ldots, x_m) - k(\xi, P)\right) P(d\xi)$$

we have

$$n^{-1/2} \sum_1^n \left(k^{(n-1)}(x_\nu, x_{n \cdot \nu}) - k(x_\nu, P)\right) \qquad (3.23)$$

$$= n^{-1/2} \sum_1^n \Delta_{n-1}(x_\nu, x_{n \cdot \nu})$$

$$+ n^{-1/2} \sum_1^n \left(\int k^{(n-1)}(\xi, x_{n \cdot \nu}) P(d\xi) - \kappa(P)\right).$$

Since (3.19) and (3.20) imply (L.15) and (L.16), we have

$$n^{-1/2} \sum_1^n \Delta_{n-1}(x_\nu, x_{n \cdot \nu}) = o_p(n^0) \qquad (P^n)$$

by Lemma L.13.

Relation (3.20) implies

$$\int \left(k^{(n)}(\xi; x_1, x_2, \ldots, x_n) - k^{(n)}(\xi; y_1, x_2, \ldots, x_n)\right)^2$$
$$P(d\xi) P(dy_1) P(dx_1) P(dx_2) \ldots P(dx_n) = o(n^{-1}),$$

hence also

$$\int \left(k^{n-1)}(\xi, x_{n \cdot \nu}) - k^{(n-1)}(\xi, x_{n \cdot 1})\right)^2 P(d\xi) P(dx_1) \ldots P(dx_n) = o(n^{-1}),$$

and therefore

$$n^{-1/2} \sum_{\nu=1}^n \int \left(k^{(n-1)}(\xi, x_{n \cdot \nu}) - k^{(n-1)}(\xi, x_{n \cdot 1})\right) P(d\xi) = o_p(n^{-1/2}) \quad (P^n).$$

Together with (3.10′) this implies

$$n^{-1/2} \sum_{\nu=1}^n \left(\int k^{(n-1)}(\xi, x_{n \cdot \nu}) P(d\xi) - \kappa(P)\right) = o_P(n^0) \quad (P^n).$$

Hence (3.21) follows from (3.23).

$\square$

There is one particular situation in which the difficulties discussed above disappear: The gradient K depends on P only through a finite dimensional functional. To work with such a restricted class of gradients will, probably, suffice for most practical purposes.

To be more specific, we presume that $k(x, P) = \underline{k}(x, \alpha(P))$, with $\underline{k} : X \times \mathbb{R}^{m+1} \to \mathbb{R}$ and $\alpha : \mathcal{P} \to \mathbb{R}^{m+1}$. We write $\alpha(P) = (\alpha_0(P), \alpha_1(P), \ldots, \alpha_m(P))$ and assume w.l.g. that $\alpha_0(P) = \kappa(P)$.

For any function $h(\alpha_0, \alpha_1, \ldots, \alpha_m)$ we denote

$$h^{(i)}(\alpha_0, \alpha_1, \ldots, \alpha_m) := \frac{\partial}{\partial \alpha_i} h(\alpha_0, \alpha_1, \ldots, \alpha_m).$$

Assume that $\underline{k}(x, \cdot)$ is differentiable for every $x \in X$, and each functional α_i is differentiable with canonical gradient $\alpha_i^*(\cdot, P)$. Then $P \to k(x, P)$ is differentiable with canonical gradient

$$\hat{k}(x, y, P) = \sum_0^m \underline{k}^{(i)}(x, \alpha(P)) \alpha_i^*(y, P). \qquad (3.24)$$

Hence

$$\int \hat{k}(x, y, P) P(dx) = \sum_0^m P(\underline{k}^{(i)}(\cdot, \alpha(P))) \alpha_i^*(y, P). \qquad (3.25)$$

If the canonical gradients $\alpha_0^*(\cdot, P)$, $\alpha_1^*(\cdot, P), \ldots, \alpha_m^*(\cdot, P)$ are linearly independent, relation (3.8) is equivalent to

$$P(\underline{k}^{(i)}(\cdot, \alpha(P))) = 0 \qquad \text{for } i = 0, 1, \ldots, m. \qquad (3.26)$$

Written in terms of the gradient K, with $\alpha_0(P) = \kappa(P)$, relation (3.26) becomes

$$P(\underline{K}^{(i)}(\cdot, \alpha(P))) = -\delta_{0i}, \qquad i = 0, 1, \ldots, m. \qquad (3.27)$$

Relation (3.26) was obtained under the assumption that the function $\underline{K}(\cdot, \alpha)$ is a gradient for κ at P if $\alpha = \alpha(P)$. In the following sections we shall use whole families of gradients $\underline{K}(\cdot, \alpha)$, where at least some of the α_i are allowed to vary freely. Of course, relation (3.26) also holds for these components.

Proposition 3.28. Assume that (3.26) holds, and that the functions $\underline{K}^{(i)}$, $i = 0, 1, \ldots, m$, fulfill condition $L.5(\alpha(P), P)$.

Then the e.s. $\kappa^{(n)}(\underline{x}) := n^{-1} \sum_{1}^{n} \underline{k}(x_\nu, \alpha^{(n)}(\underline{x}))$, $n \in \mathbb{N}$, is as. linear with gradient $\underline{K}$, provided $\alpha^{(n)} = \alpha(P) + O_p(n^{-1/2})$ (P^n).

Addendum. $\alpha^{(n)} = \alpha(P) + o_p(n^{-1/4})$ suffices if for all $i = 0, 1, \ldots, m$, $\underline{K}^{(i)}(\cdot, \alpha(P))$ is P–square integrable, and if $\alpha \to \underline{K}^{(i)}(\cdot, \alpha)$ fulfill at $\alpha = \alpha(P)$ a Lipschitz–condition of the following type.

There exists a P-integrable function $c : X \to \mathbb{R}$ such that

$$|\underline{K}^{(i)}(x, \alpha) - \underline{K}^{(i)}(x, \alpha(P))| \leq \|\alpha - \alpha(P)\| c(x) \tag{3.29}$$

for all α in a neighborhood of $\alpha(P)$.

Proof. Condition (3.4) follows immediately from Proposition L.10, applied with $\underline{k}(x, \alpha_0, \alpha_1, \ldots, \alpha_m)$ in place of $h(x, \vartheta_1, \ldots, \vartheta_m)$.

To prove condition (3.4) under the assumptions of the Addendum, we use the Taylor expansion

$$\sum_{1}^{n} \left(\underline{k}(x_\nu, \alpha^{(n)}(\underline{x})) - \underline{k}(x_\nu, \alpha(P)) \right)$$

$$= \sum_{i=0}^{m} (\alpha_i^{(n)}(\underline{x}) - \alpha_i(P)) \sum_{\nu=1}^{n} \underline{k}^{(i)}(x_\nu, \alpha(P)) + \sum_{0}^{m} R_i^{(n)}(\underline{x}, P),$$

with

$$R_i^{(n)}(\underline{x}, P) = (\alpha_i^{(n)}(\underline{x}) - \alpha_i(P))$$
$$\sum_{\nu=1}^{n} \int_0^1 [\underline{k}^{(i)}(x_\nu, (1-u)\alpha(P) + u\alpha^{(n)}(\underline{x})) - \underline{k}^{(i)}(x_\nu, \alpha(P))] \, du.$$

Since

$$|R_i^{(n)}(\underline{x}, P)| \leq \|\alpha^{(n)}(\underline{x}) - \alpha(P)\|^2 \sum_{1}^{n} c(x_\nu),$$

$\alpha^{(n)} = \alpha(P) + o_p(n^{-1/4})$ (P^n) implies

$$n^{-1/2} \sum_{0}^{m} R_i^{(n)}(\underline{x}, P) = o_p(n^0) \quad (P^n).$$

Since $\underline{K}^{(i)}(\cdot, \alpha(P))$ is P–square integrable, $n^{-1/2} \sum_{\nu=1}^{n} \underline{k}^{(i)}(x_\nu, \alpha(P))$ is stochastically bounded because of (3.26). Hence

$$n^{-1/2} \sum_{i=1}^{m} (\alpha_i^{(m)}(\underline{x}) - \alpha_i(P)) \sum_{\nu=1}^{n} \underline{k}^{(i)}(x_\nu, \alpha(P)) = o_p(n^0)$$

whenever $\alpha^{(n)} = \alpha(P) + o_p(n^0)$.

□

4. Estimation in semiparametric models

In this section we turn to a more special model, $\mathcal{P} = \{P_{\vartheta,\tau} : \vartheta \in \Theta, \tau \in T\}$, with $\Theta \subset \mathbb{R}$, and T some set (endowed with a topology, if necessary). The problem is to estimate ϑ, with τ unknown, i.e. we are dealing with the functional $\kappa(P_{\vartheta,\tau}) = \vartheta$. T can be a set of functions, a set of measures, and — of course — also a subset of a Euclidean space. Thanks to this more special model, more precise instructions for the construction of estimators can be given.

For convenience of notation, let $f^\bullet$ denote the derivative with respect to ϑ, for any function f.

Let $T_0(P_{\vartheta,\tau})$ denote the tangent space of $\{P_{\vartheta,\tilde{\tau}} : \tilde{\tau} \in T\}$ at $P_{\vartheta,\tau}$. Intuitively speaking, $T_0(P_{\vartheta,\tau})$ consists of all those directions in the tangent space $T(P_{\vartheta,\tau}, \mathcal{P})$ in which the functional is "locally constant" (i.e. $\lim_{t \to 0} t^{-1}(\kappa(P_t) - \kappa(P)) = 0$ for all paths P_t, $t \to 0$, converging to P from a direction in $T_0(P_{\vartheta,\tau})$). This interpretation justifies the name of "level space" for $T_0(P_{\vartheta,\tau})$. (More generally, $T_0(P_{\vartheta,\tau})$ is the orthogonal complement in $T(P_{\vartheta,\tau}, \mathcal{P})$ of the canonical gradient.)

Throughout the following we assume that

$$T(P_{\vartheta,\tau}, \mathcal{P}) = \{c\ell^\bullet(\cdot, \vartheta, \tau) + h : c \in \mathbb{R}, h \in T_0(P_{\vartheta,\tau})\}, \tag{4.1}$$

i.e. that the direction of any differentiable path $P_{\vartheta_t, \tau_t} \to P_{\vartheta,\tau}$ is made up additively of the directions originating from the two paths $P_{\vartheta_t, \tau} \to P_{\vartheta,\tau}$ and $P_{\vartheta, \tau_t} \to P_{\vartheta,\tau}$.

Since $\mathcal{P}$ remains fixed, we write $T(P_{\vartheta,\tau})$ for $T(P_{\vartheta,\tau}, \mathcal{P})$.

Considering paths $P_{\vartheta+ct, \tau_t}$ converging to $P_{\vartheta,\tau}$ from direction $c\ell^\bullet(\cdot, \vartheta, \tau) + h$ we find that $K(\cdot, \vartheta, \tau) \in \mathcal{L}_*(P_{\vartheta,\tau})$ is a gradient iff (see (1.3)) for all $c \in \mathbb{R}$, $h \in T_0(P_{\vartheta,\tau})$

$$\lim_{t \to 0} t^{-1}\big(\kappa(P_{\vartheta+ct,\tau_t}) - \kappa(P_{\vartheta,\tau})\big)$$
$$= P\big(K(\cdot, \vartheta, \tau)(c\ell^\bullet(\cdot, \vartheta, \tau) + h)\big).$$

Since the left-hand side equals c, we obtain the following conditions for K

$$P_{\vartheta,\tau}\big(K(\cdot, \vartheta, \tau)\ell^\bullet(\cdot, \vartheta, \tau)\big) = 1 \tag{4.2}$$
$$P_{\vartheta,\tau}\big(K(\cdot, \vartheta, \tau)h\big) = 0 \quad \text{for } h \in T_0(P_{\vartheta,\tau}). \tag{4.3}$$

(It would be more consistent with the notation of Sections 1 and 2 to denote the gradients by $\kappa^+(\cdot, P_{\vartheta,\tau})$ rather than $K(\cdot, \vartheta, \tau)$. But we need derivatives with respect to ϑ, and $\kappa^{+\bullet}(\cdot, P_{\vartheta,\tau})$ does not look nice.)

Let $L(\cdot,\vartheta,\tau)$ denote the orthogonal component of $\ell^\bullet(\cdot,\vartheta,\tau)$ with respect to $T_0(P_{\vartheta,\tau})$.

Since any gradient $K(\cdot,\vartheta,\tau)$ is by (4.3) orthogonal to $T_0(P_{\vartheta,\tau})$, we have

$$P_{\vartheta,\tau}\bigl(K(\cdot,\vartheta,\tau)L(\cdot,\vartheta,\tau)\bigr) = P_{\vartheta,\tau}\bigl(K(\cdot,\vartheta,\tau)\ell^\bullet(\cdot,\vartheta,\tau)\bigr) \qquad (4.4)$$

and (4.2) may be rewritten as

$$P_{\vartheta,\tau}\bigl(K(\cdot,\vartheta,\tau)L(\cdot,\vartheta,\tau)\bigr) = 1. \qquad (4.5)$$

Throughout the following we presume that the order between differentiation and integration may be interchanged. Under this condition, there is another condition equivalent to (4.2), namely

$$P_{\vartheta,\tau}\bigl(K^\bullet(\cdot,\vartheta,\tau)\bigr) = -1. \qquad (4.6)$$

Since $T(P_{\vartheta,\tau})$ is the linear space spanned by $\ell^\bullet(\cdot,\vartheta,\tau)$ and $T_0(P_{\vartheta,\tau})$, we obtain from (4.3) that $\kappa^*(\cdot,\vartheta,\tau)$, the canonical gradient of the functional $\kappa(P_{\vartheta,\tau}) = \vartheta$, is proportional to $L(\cdot,\vartheta,\tau)$. Condition (4.5) implies that

$$\kappa^*(x,\vartheta,\tau) = L(x,\vartheta,\tau)/P_{\vartheta,\tau}\bigl(L(\cdot,\vartheta,\tau)^2\bigr). \qquad (4.7)$$

According to Section 2, the as. variance bound is

$$P_{\vartheta,\tau}\bigl(\kappa^*(\cdot,\vartheta,\tau)^2\bigr) = 1/P_{\vartheta,\tau}\bigl(L(\cdot,\vartheta,\tau)^2\bigr). \qquad (4.8)$$

For unbiased estimators for ϑ, $1/P_{\vartheta,\tau}\bigl(L(\cdot,\vartheta,\tau)^2\bigr)$ admits a finite–sample–size interpretation as minimal variance. This is a straightforward generalization of the Cramér–Rao bound.

Remark 4.9. If $\ell^\bullet(\cdot,\vartheta,\tau)$ itself is orthogonal to $T_0(P_{\vartheta,\tau})$, we have $L(\cdot,\vartheta,\tau) = \ell^\bullet(\cdot,\vartheta,\tau)$ and the as. variance bound (4.8) becomes $1/P_{\vartheta,\tau}\bigl(\ell^\bullet(\cdot,\vartheta,\tau)^2\bigr)$. This is the celebrated case of "adaptivity", in which the as. variance bound is the same, irrespectively of whether or not τ is known.

Concerning the existence of e.s. with a given influence function K, not much more can be said than in Proposition 3.12. Since a preliminary e.s. for ϑ is needed anyway, it now appears natural to seek first an estimator $\underline{x} \to K^{(n)}(\cdot,\vartheta,\underline{x})$ for $K(\cdot,\vartheta,\tau)$ and to replace ϑ by the preliminary estimator $\vartheta^{(n)}$ afterwards. For this purpose, conditions (3.10) can be given in a slightly different form. Switching from $k^{(n)}(\cdot,\vartheta,\underline{x}) = \vartheta + K^{(n)}(\cdot,\vartheta,\underline{x})$ to $K^{(n)}(\cdot,\vartheta,\underline{x})$ they can be written as follows.

For any (nonstochastic) sequence $\vartheta_n = \vartheta_0 + O(n^{-1/2})$,

$$\int K^{(n)}(x, \vartheta_n, \underline{x}) P_{\vartheta_n, \tau_0}(dx) = o_p(n^{-1/2}) \quad (P^n_{\vartheta_n, \tau_0}), \tag{4.10'}$$

$$\int \left(K^{(n)}(x, \vartheta_n, \underline{x}) - K(x, \vartheta_n, \tau_0) \right)^2 P_{\vartheta_n, \tau_0}(dx) = o_p(n^0) \quad (P^n_{\vartheta_n, \tau_0}). \tag{4.10''}$$

In this form, these conditions appear in Schick (1986, p. 1141, (2.3) and (2.4)) and Klaassen (1987, p. 1549, (1.5) and (1.4)) with $K(\cdot, \vartheta, \tau)$ being the canonical gradient. (The emphasis of Klaassen's paper is that these conditions are even necessary, in a certain sense.)

Using the splitting technique and a discretized version of $\vartheta^{(n)}$, Schick obtains as. efficiency of an appropriately defined e.s. Refining the art of splitting, Klaassen obtains this result without discretization. He argues as follows. Relations (4.10) can be used in the same way as (3.10) to prove that

$$m_n^{-1} \sum_{1}^{m_n} \left(k^{(n-\ell_n)}(x_\nu, \vartheta_n; x_{\ell_n+1}, \ldots, x_n) \right. \tag{4.11}$$
$$\left. - k(x_\nu, \vartheta_n, \tau_0) \right) = o_p(n^0) \quad (P^n_{\vartheta_n, \tau_0}),$$

with m_n/n, $(\ell_n - m_n)/n$ and $(n - \ell_n)/n$, $n \in \mathbb{N}$, bounded away from 0. Under mild regularity conditions on the family $\{P_{\vartheta, \tau_0} : \vartheta \in \Theta\}$, the condition $\vartheta_n = \vartheta_0 + O(n^{-1/2})$ implies that $P^n_{\vartheta_n, \tau_0}$ and $P^n_{\vartheta_0, \tau_0}$ are contiguous, so that (4.11) holds with a remainder term $o_p(n^0)$ $(P^n_{\vartheta_0, \tau_0})$. By Corollary L.11,

$$m_n^{-1} \sum_{1}^{m_n} \left(k(x_\nu, \vartheta_n, \tau_0) - k(x_\nu, \vartheta_0, \tau_0) \right) = o_p(n^0) \quad (P^n_{\vartheta_0, \tau_0})$$

if $(x, \vartheta) \to k^\bullet(x, \vartheta, \tau_0)$ fulfills condition L.5$(\vartheta_0, P_{\vartheta_0, \tau_0})$. Hence (4.11) implies

$$m_n^{-1} \sum_{1}^{m_n} \left(k^{(\ell_n - m_n)}(x_\nu, \vartheta_n; x_{m_n+1}, \ldots, x_{\ell_n}) - k(x_\nu, \vartheta_0, \tau_0) \right) \tag{4.12}$$
$$= o_p(n^0) \quad (P^n_{\vartheta_0, \tau_0}).$$

Since $x_{\ell_n+1}, \ldots, x_n$ are stochastically independent of $x_1, \ldots, x_{\ell_n}$ occurring in (4.12), the nonstochastic sequence $\vartheta_n = \vartheta_0 + O(n^{-1/2})$ may be replaced by the stochastic sequence $\vartheta^{(n-\ell_n)}(x_{\ell_n+1}, \ldots, x_n) = \vartheta_0 + O_p(n^{-1/2})$, by Lemma L.19.

Therefore, relation (3.11) holds with $k^{(n-m_n)}(x; x_{m_n+1}, \ldots, x_n)$ replaced by $k^{(\ell_n - m_n)}(x, \vartheta^{(n-\ell_n)}(x_{\ell_n+1}, \ldots, x_n); x_{m_n+1}, \ldots, x_{\ell_n})$, and the existence of an e.s. with influence function $K(\cdot, P)$ follows from Proposition 3.12.

As a consequence of Proposition 3.6 we obtain that there always exist e.s. $K^{(n)}(\cdot, \vartheta, \underline{x})$ for $K(\cdot, \vartheta, \tau)$ of the type $K(x, \vartheta, \tau^{(n)}(\underline{x}))$. Restricting the attention to such e.s. from the beginning opens the possibility of using the regularity of $\vartheta \to K(x, \vartheta, \tau)$. More precisely: If $K^\bullet$ fulfills condition L.5$((\vartheta_0, \tau_0), P_{\vartheta_0, \tau_0})$, then the same holds true of $k^\bullet$. Using $P_{\vartheta_0, \tau_0}(k^\bullet(\cdot, \vartheta_0, \tau_0)) = 0$ we obtain from Proposition L.10 that

$$n^{-1/2} \sum_1^n \left(k(x_\nu, \vartheta^{(n)}(\underline{x}), \tau^{(n)}(\underline{x})) - k(x_\nu, \vartheta_0, \tau^{(n)}(\underline{x})) \right) = o_p(n^0) \quad (P^n_{\vartheta_0, \tau_0}),$$

provided the e.s. fulfill $\vartheta^{(n)} = \vartheta_0 + O_p(n^{-1/2})$ and $\tau^{(n)} = \tau_0 + o_p(n^0)$. Hence condition (3.4) will be fulfilled with $k^{(n)}(x, \underline{x}) = k(x, \vartheta^{(n)}(\underline{x}), \tau^{(n)}(\underline{x}))$ iff

$$n^{-1/2} \sum_1^n \left(k(x_\nu, \vartheta_0, \tau^{(n)}(\underline{x})) - k(x_\nu, \vartheta_0, \tau_0) \right) = o_p(n^0) \quad (P^n_{\vartheta_0, \tau_0}),$$

which is the same as

$$n^{-1/2} \sum_1^n \left(K(x_\nu, \vartheta_0, \tau^{(n)}(\underline{x})) - K(x_\nu, \vartheta_0, \tau_0) \right) = o_p(n^0) \quad (P^n_{\vartheta_0, \tau_0}). \tag{4.13}$$

If we circumvent the dependence between x_ν and $\tau^{(n)}(\underline{x})$ by splitting, conditions (4.10) with ϑ_n replaced by ϑ_0 suffice. Among these, condition (4.10') is the difficult one. It requires that

$$\int K(x, \vartheta_0, \tau^{(n)}(\underline{x})) P_{\vartheta_0, \tau_0}(dx) = o_p(n^{-1/2}) \quad (P^n_{\vartheta_0, \tau_0}) \tag{4.14}$$

which will, in general, not be true unless $\tau^{(n)}$ converges to τ_0 at a certain rate.

* *

*

For the purpose of illustration, we consider now the case of a parametric family $\mathcal{P} = \{P_{\vartheta, \tau} : \vartheta \in \Theta, \tau \in \mathrm{T}\}$ with $\Theta \subset \mathbb{R}$ and $\mathrm{T} \subset \mathbb{R}^m$. Let

$$\ell^{(0)}(\cdot, \vartheta, \tau) = \frac{\partial}{\partial \vartheta} \log p(\cdot, \vartheta, \tau), \quad \ell^{(i)}(\cdot, \vartheta, \tau) = \frac{\partial}{\partial \tau_i} \log p(\cdot, \vartheta, \tau)$$

for $i = 1, \ldots, m$. Moreover, we need the matrix $L(\vartheta, \tau)$ with components

$$L_{i,j}(\vartheta, \tau) = \int \ell^{(i)}(\cdot, \vartheta, \tau) \ell^{(j)}(\cdot, \vartheta, \tau) dP_{\vartheta, \tau}, \quad i, j = 0, 1, \ldots, m,$$

and its inverse $\Lambda(\vartheta,\tau)$. We presume the usual conditions

$$\int \ell^{(i)}(\cdot,\vartheta,\tau)dP_{\vartheta,\tau} = 0 \quad \text{and} \quad \int \ell^{(ij)}(\cdot,\vartheta,\tau)dP_{\vartheta,\tau} = -L_{i,j}(\vartheta,\tau)$$

for $i,j = 0,1,\ldots,m$.

Under suitable regularity conditions on the densities (see I, p. 35, Proposition 2.2.1) we obtain

$$\text{tangent space } T(P_{\vartheta,\tau}) = \{\sum_{0}^{m} a_i \ell^{(i)}(\cdot,\vartheta,\tau) : a_i \in \mathbb{R}, \ i = 0,1,\ldots,m\}, \quad (4.15)$$

$$\text{level space } \quad T_0(P_{\vartheta,\tau}) = \{\sum_{1}^{m} a_i \ell^{(i)}(\cdot,\vartheta,\tau) : a_i \in \mathbb{R}, \ i = 1,\ldots,m\}. \quad (4.16)$$

The orthogonal component of $\ell^{(0)}(\cdot,\vartheta,\tau)$ with respect to $T_0(P_{\vartheta,\tau})$ is

$$L(\cdot,\vartheta,\tau) = \sum_{0}^{m} \frac{\Lambda_{0j}(\vartheta,\tau)}{\Lambda_{00}(\vartheta,\tau)} \ell^{(j)}(\cdot,\vartheta,\tau), \quad (4.17)$$

the canonical gradient is

$$\kappa^*(\cdot,\vartheta,\tau) = \sum_{0}^{m} \Lambda_{0j}(\vartheta,\tau) \ell^{(j)}(\cdot,\vartheta,\tau). \quad (4.18)$$

Proposition 4.19. *Assume that the functions* $(x,\vartheta,\tau) \to \ell^{(ij)}(x,\vartheta,\tau)$ *fulfill condition* L.5$((\vartheta_0,\tau_0), P_{\vartheta_0,\tau_0})$. *If* $\vartheta^{(n)} = \vartheta_0 + O_p(n^{-1/2})$, $\tau^{(n)} = \tau_0 + O_p(n^{-1/2})$, $a_j^{(n)} = \Lambda_{0j}(\vartheta_0,\tau_0)/\Lambda_{00}(\vartheta_0,\tau_0) + o_p(n^0)$, $j = 0,\ldots,m$, *with respect to* P_{ϑ_0,τ_0}^n, *then the e.s.*

$$\hat{\vartheta}^{(n)}(\underline{x}) := \vartheta^{(n)}(\underline{x}) + n^{-1} \sum_{j=0}^{m} a_j^{(n)}(\underline{x}) \sum_{\nu=1}^{n} \ell^{(j)}(x_\nu, \vartheta^{(n)}(\underline{x}), \tau^{(n)}(\underline{x})) \quad (4.20)$$

is as. efficient.

Addendum. *If the functions* $\ell^{(ij)}$, $i,j = 0,1,\ldots,m$, *fulfill a Lipschitz–condition of the type* (3.29), *then* $O_p(n^{-1/2})$ *may be replaced by* $o_p(n^{-1/4})$ *in the conditions for* $\vartheta^{(n)}$ *and* $\tau^{(n)}$.

Proof. Follow the proof of Proposition 3.28, taking into account that the function $a \to k(x,\vartheta,\tau,a) := \vartheta + \sum_{0}^{m} a_j \ell^{(j)}(x,\vartheta,\tau)$ is linear.

□

Remark 4.21. According to Proposition 4.19, the conditions on the preliminary e.s. for the nuisance parameter τ are rather weak. This results from the fact that any gradient for the functional $\kappa(P_{\vartheta,\tau}) = \vartheta$ is orthogonal to the level space. For this reason the canonical gradient (being proportional to the *orthogonal component* of $\ell^{\bullet}$ with respect to the level space) is easier to estimate than $\ell^{\bullet}$ itself. (Without this property, estimation in semiparametric models would be hopeless.)

For parametric families this means that $\kappa^*(\cdot, \vartheta, \tau)$, given by (4.18), is easier to estimate than $\ell^{(0)}(\cdot, \vartheta, \tau)$, because

$$\int \frac{\partial}{\partial \tau_i} \kappa^*(\cdot, \vartheta, \tau) dP_{\vartheta,\tau} = 0 \qquad \text{for } i = 1, \ldots, m.$$

It appears that this fact has been overlooked by some authors.

The weak conditions on $\tau^{(n)}$ required in Proposition 4.19 seem also to be in contrast to Gong and Samaniego (1981), who treat the following problem. If the m.l. estimator is difficult to obtain, they suggest to substitute a preliminary estimator $\tau^{(n)}$ for the nuisance parameter, and to solve

$$\sum_{1}^{n} \ell^{(0)}\left(x_\nu, \vartheta, \tau^{(n)}(\underline{x})\right) = 0 \qquad (4.22)$$

for ϑ. Gong and Samaniego obtain the as. distribution of the resulting estimator, which is, in general, as. efficient for ϑ if and only if $\tau^{(n)}$ is as. efficient for τ.

That they need an as. efficient e.s. for τ results from the fact that (4.22) is the obvious estimating equation, but not the most well considered one. Theory tells us that (see (4.18))

$$\sum_{\nu=1}^{n} \sum_{j=0}^{m} \Lambda_{0j}\left(\vartheta, \tau^{(n)}(\underline{x})\right) \ell^{(j)}\left(x_\nu, \vartheta, \tau^{(n)}(\underline{x})\right) = 0$$

is preferable. With this estimating equation, a $\sqrt{n}$-consistent $\tau^{(n)}$ is sufficient (and even $\tau^{(n)} = \tau_0 + o_p(n^{-1/4})$ is enough under the Lipschitz-condition for $\ell^{(ij)}$).

5. Families of gradients

In this section we turn to the following situation:

For $(\vartheta, \tau) \in \Theta \times T$ (with $\Theta \subset \mathbb{R}$ and T a general parameter space) we are given a family of functions $\{N(\cdot, \vartheta, \alpha) : \alpha \in A_\tau\}$ in $\mathcal{L}_*(P_{\vartheta,\tau})$ with the following property:

For any $\alpha \in A_\tau$, $N(\cdot, \vartheta, \alpha)$ is orthogonal to $T_0(P_{\vartheta,\tau})$. (5.1)

This includes two important special cases

a) $A_\tau = \{\tau\}$ for $\tau \in T$

b) $A_\tau = A$ for $\tau \in T$.

A natural situation leading to (5.1.b) is

$$P_{\vartheta,\tau}(N(\cdot, \vartheta, \alpha)) = 0 \quad \text{for all } \alpha \in A. \tag{5.2}$$

Applied with τ replaced by a path $\tau_t \to \tau$, (5.2) implies (under appropriate conditions on the remainder term in this path, say $P(r_t^2) = o(t^0)$) that $N(\cdot, \vartheta, \alpha)$ is orthogonal to $T_0(P_{\vartheta,\tau})$.

Any function N fulfilling (5.1) can be transformed into a gradient by standardization. According to (4.2), $c(\vartheta, \tau, \alpha) N(\cdot, \vartheta, \alpha)$ is a gradient for $\kappa(\vartheta, \tau) = \vartheta$ iff

$$c(\vartheta, \tau, \alpha) = 1/P_{\vartheta,\tau}(N(\cdot, \vartheta, \alpha)\ell^\bullet(\cdot, \vartheta, \tau)). \tag{5.3}$$

If the order between differentiation with respect to ϑ and integration can be interchanged, we obtain from $P_{\vartheta,\tau}(N(\cdot, \vartheta, \alpha)) = 0$ for $\vartheta \in \Theta$ that

$$P_{\vartheta,\tau}(N(\cdot, \vartheta, \alpha)\ell^\bullet(\cdot, \vartheta, \tau)) = -P_{\vartheta,\tau}(N^\bullet(\cdot, \vartheta, \alpha)). \tag{5.4}$$

Hence, for every $\alpha \in A$,

$$K(x, \vartheta, \tau, \alpha) := -N(x, \vartheta, \alpha)/P_{\vartheta,\tau}(N^\bullet(\cdot, \vartheta, \alpha)) \tag{5.5}$$

is a gradient (at $P_{\vartheta,\tau}$) for the functional $\kappa(\vartheta, \tau) = \vartheta$.

Theorem 5.6 below refers to arbitrary functions N fulfilling (5.1). The main subject of this section is, however, the more special case described by (5.2).

To use families of functions N fulfilling (5.2) has certain advantages, provided one is ready to settle with as. (slightly) inefficent e.s.

(i) It is not necessary to estimate the nuisance parameter τ. The latter occurs in the gradient given by (5.5) through the $P_{\vartheta,\tau}$-integral only, which can be estimated by the sample mean.

(ii) To obtain e.s. for $N(\cdot,\vartheta,\alpha)$ fulfilling condition 5.6(iv) requires, in general, e.s. converging to α at a certain rate, hence a structure on A more special than a topology. Continuity of $\alpha \to N(\cdot,\vartheta,\alpha)$ alone is not enough to guarantee condition 5.6(iv) for all e.s. $\alpha^{(n)} = \alpha_0 + o_p(n^0)$. Though it is impossible to establish condition 5.6(iv) for an arbitrary e.s. $\alpha^{(n)}$ just on the basis of $\alpha^{(n)} = \alpha_0 + o_p(n^0)$, it is, at the same time, possible to improve this e.s. to enforce 5.6(iv). This will be shown in Proposition 5.12 below.

(iii) If one insists on as. efficient e.s. for ϑ, the regularity conditions stated in Theorem 5.6 have to be fulfilled with $N(\cdot,\vartheta,\alpha)$ replaced by $L(\cdot,\vartheta,\tau)$. Confining oneself to as. subefficient e.s. opens the possibility of choosing a regular family $N(\cdot,\vartheta,\alpha)$ (thus reducing the regularity conditions on the family of p-measures).

(iv) Given a family of functions $N(\cdot,\vartheta,\alpha)$, $\alpha \in A$, fulfilling (5.2), it is hard to think of any application where it would not suffice to consider a finite dimensional subfamily of A. In this case, all the difficulties connected with condition 5.6(iv) disappear. For parametric families it is usually easy to obtain e.s. $\alpha^{(n)} = \alpha_0 + O_p(n^{-1/2})$, and this suffices to establish condition 5.6(iv). If $\alpha \to N(\cdot,\vartheta,\alpha)$ is sufficiently regular, even $\alpha^{(n)} = \alpha_0 + o_p(n^{-1/4})$ suffices (see Propositions 3.28 and 4.19).

To illustrate the simplification achievable in this way, we refer to Example 1, where any function $N((x_1,x_2),\vartheta,\alpha) = (\vartheta x_1 - x_2)/(\alpha + \vartheta x_1 + x_2)$, $\alpha \in \mathbb{R}_+$, is orthogonal to $T_0(P_{\vartheta,\Gamma})$. The canonical gradient is $L(\cdot,\vartheta,\Gamma)/P_{\vartheta,\Gamma}(L(\cdot,\vartheta,\Gamma)^2)$, with

$$L((x_1,x_2),\vartheta,\Gamma) = -\frac{\vartheta x_1 - x_2}{2\vartheta^{3/2}} \frac{\int \eta^3 \exp[-\eta\vartheta^{-1/2}(\vartheta x_1 + x_2)]\Gamma(d\eta)}{\int \eta^2 \exp[-\eta\vartheta^{-1/2}(\vartheta x_1 + x_2)]\Gamma(d\eta)}.$$

Throughout the following we suppose that the set A_τ is endowed with a Hausdorff topology with countable base.

The following theorem provides a general framework for subsequent applications. It requires no e.s. $\alpha^{(n)}$ and no assumption like (5.2). Such assumptions, however, are required to establish Condition 5.6(iv).

Let $\alpha_0 \in A_{\tau_0}$ be an arbitrary element. Throughout the following we presume that $P_{\vartheta_0,\tau_0}(N(\cdot,\vartheta_0,\alpha_0)^2) > 0$ and $P_{\vartheta_0,\tau_0}(N^\bullet(\cdot,\vartheta_0,\alpha_0)) \neq 0$.

Theorem 5.6. *Assume the following conditions.*

(i) $(x,\vartheta) \to N^\bullet(x,\vartheta,\alpha_0)$ fulfills L.5($\vartheta_0, P_{\vartheta_0,\tau_0}$),

(ii) $\vartheta^{(n)} = \vartheta_0 + O_p(n^{-1/2})$ $(P^n_{\vartheta_0,\tau_0})$,

(iii) $d^{(n)} = P_{\vartheta_0,\tau_0}(N^\bullet(\cdot,\vartheta_0,\alpha_0)) + o_p(n^0)$ $(P^n_{\vartheta_0,\tau_0})$.

(iv) There exist maps $N_\nu^{(n)} : \Theta \times X^n \to \mathbb{R}$, $\nu = 1, \ldots, n$, such that

$$n^{-1/2} \sum_{1}^{n} \left(N_\nu^{(n)}(\vartheta^{(n)}(\underline{x}), \underline{x}) - N(x_\nu, \vartheta^{(n)}(\underline{x}), \alpha_0) \right) = o_p(n^0) \quad (P_{\vartheta_0, \tau_0}^n).$$

Then the e.s. defined by

$$\hat{\vartheta}^{(n)}(\underline{x}) := \vartheta^{(n)}(\underline{x}) - n^{-1} \sum_{1}^{n} N_\nu^{(n)}\left(\vartheta^{(n)}(\underline{x}), \underline{x}\right) / d^{(n)}(\underline{x}) \tag{5.7}$$

is under $P_{\vartheta_0, \tau_0}^n$ as. normal with mean 0 and variance

$$\sigma^2(\vartheta_0, \tau_0, \alpha_0) = P_{\vartheta_0, \tau_0}\left(N(\cdot, \vartheta_0, \alpha_0)^2\right) / \left(P_{\vartheta_0, \tau_0}(N^\bullet(\cdot, \vartheta_0, \alpha_0))\right)^2. \tag{5.8}$$

Here and throughout the following we use the phrase "$\vartheta^{(n)}$, $n \in \mathbb{N}$, is as. normal with mean 0 and variance σ^2" as an abbreviation for "the standardized e.s. $n^{1/2}(\vartheta^{(n)} - \vartheta_0)$, $n \in \mathbb{N}$, is as. normal with mean 0 and variance σ^2".

Proof. By assumption (iv) and Proposition L.10

$$n^{-1/2} \sum_{1}^{n} \left(N_\nu^{(n)}\left(\vartheta^{(n)}(\underline{x}), \underline{x}\right) - N(x_\nu, \vartheta_0, \alpha_0) \right) \tag{5.9}$$

$$= n^{-1/2} \sum_{1}^{n} \left(N(x_\nu, \vartheta^{(n)}(\underline{x}), \alpha_0) - N(x_\nu, \vartheta_0, \alpha_0) \right) + o_p(n^0)$$

$$= n^{1/2} \left(\vartheta^{(n)}(\underline{x}) - \vartheta_0 \right) P_{\vartheta_0, \tau_0}(N^\bullet(\cdot, \vartheta_0, \alpha_0)) + o_p(n^0) \quad (P_{\vartheta_0, \tau_0}^n).$$

From (5.7), (5.9) and (iii)

$$n^{1/2}\left(\hat{\vartheta}^{(n)}(\underline{x}) - \vartheta_0\right)$$

$$= n^{1/2}\left(\vartheta^{(n)}(\underline{x}) - \vartheta_0\right) - n^{-1/2} \sum_{1}^{n} N_\nu^{(n)}\left(\vartheta^{(n)}(\underline{x}), \underline{x}\right) / d^{(n)}(\underline{x})$$

$$= -n^{-1/2} \sum_{1}^{n} N(x_\nu, \vartheta_0, \alpha_0) / P_{\vartheta_0, \tau_0}(N^\bullet(\cdot, \vartheta_0, \alpha_0)) + o_p(n^0) \quad (P_{\vartheta_0, \tau_0}^n).$$

From this the assertion follows immediately. □

Remark 5.10. So far, the e.s. $d^{(n)}$ for $P_{\vartheta_0, \tau_0}(N^\bullet(\cdot, \vartheta_0, \alpha_0))$ occurring in Theorem 5.6 under (iii) was left unspecified, because there are several candidates the suitability of which depends on the particular situation.

Assume we are given any function $(x, \vartheta, \alpha) \to N_0(x, \vartheta, \alpha)$ fulfilling condition L.5$((\vartheta_0, \alpha_0), P_{\vartheta_0, \tau_0})$ such that

$$P_{\vartheta_0, \tau_0}(N_0(\cdot, \vartheta_0, \alpha_0)) = P_{\vartheta_0, \tau_0}(N^\bullet(\cdot, \vartheta_0, \alpha_0)). \tag{5.11}$$

Then $d^{(n)}(\underline{x}) := n^{-1} \sum_{1}^{n} N_0(x_\nu, \vartheta^{(n)}(\underline{x}), \alpha^{(n)}(\underline{x}))$ fulfills condition (iii) by Proposition L.9.

If $A \equiv T$ and $N(\cdot, \vartheta, \tau) = L(\cdot, \vartheta, \tau)$, then relation (5.11) holds true with $N_0(\cdot, \vartheta, \tau) = -L(\cdot, \vartheta, \tau)^2$ (hint: use (5.4) for $N = L$) which offers an alternative to $N_0(\cdot, \vartheta, \tau) = L^\bullet(\cdot, \vartheta, \tau)$.

If $N(\cdot, \vartheta, \tau)$ itself is a gradient, we have $P_{\vartheta, \tau}(N^\bullet(\cdot, \vartheta, \tau)) = -1$ and the estimator $d^{(n)}$ can be dropped.

Other possibilities will be discussed in Remark 7.19. See also p. 65.

The following proposition presumes an e.s. $\alpha^{(n)}$ converging to *some* $\alpha_0 \in A_{\tau_0}$.

Proposition 5.12. *Assume that A_{τ_0} is a separable metric space, and that, in addition to (5.2), the following conditions are fulfilled:*

(i) $(x, \vartheta, \alpha) \to N^\bullet(x, \vartheta, \alpha)$ *fulfills* L.5$((\vartheta_0, \alpha_0), P_{\vartheta_0, \tau_0})$,

(ii) $(x, \alpha) \to N(x, \vartheta_0, \alpha)^2$ *fulfills* L.5$(\alpha_0, P_{\vartheta_0, \tau_0})$,

(iii) $\vartheta^{(n)} = \vartheta_0 + O_p(n^{-1/2})$ $(P_{\vartheta_0, \tau_0}^n)$,

(iv) $\alpha^{(n)} = \alpha_0 + o_p(n^0)$ $(P_{\vartheta_0, \tau_0}^n)$.

Then there exists a permutation invariant e.s. $\hat{\alpha}^{(n)} = \alpha_0 + o_p(n^0)$ $(P_{\vartheta_0, \tau_0}^n)$ *such that 5.6(iv) is fulfilled with*

$$N_\nu^{(n)}(\vartheta, \underline{x}) := N(x_\nu, \vartheta, \hat{\alpha}^{(n)}(\underline{x})).$$

Proof. For $n \in \mathbb{N}$ let $m_n \in \{1, \ldots, n\}$ be such that $n^{-1} m_n$, $n \in \mathbb{N}$, is bounded away from 0 and 1.

Since $(x, \alpha) \to N(x, \vartheta_0, \alpha)^2$ fulfills condition L.5$(\alpha_0, P_{\vartheta_0, \tau_0})$ so does

$$(x, \alpha) \to \big(N(x, \vartheta_0, \alpha) - N(x, \vartheta_0, \alpha_0)\big)^2,$$

and we obtain from Corollary L.8 that

$$\alpha \to P_{\vartheta_0, \tau_0}\Big(\big(N(\cdot, \vartheta_0, \alpha) - N(\cdot, \vartheta_0, \alpha_0)\big)^2\Big)$$

is continuous at $\alpha = \alpha_0$. Therefore

$$\underline{x} \to P_{\vartheta_0,\tau_0}\left(\left(N(\cdot,\vartheta_0,\alpha^{(n-m_n)}(x_{m_n+1},\ldots,x_n)) - N(\cdot,\vartheta_0,\alpha_0)\right)^2\right)$$
$$= o_p(n^0) \quad (P_{\vartheta_0,\tau_0}^n).$$

Since $P_{\vartheta_0,\tau_0}\left(N(\cdot,\vartheta_0,\alpha^{(n-m_n)}(x_{m_n+1},\ldots,x_n))\right) = 0$ for all $\underline{x} \in X^n$, Lemma L.12 applies and we obtain

$$m_n^{-1/2} \sum_1^{m_n} [N(x_\nu,\vartheta_0,\alpha^{(n-m_n)}(x_{m_n+1},\ldots,x_n)) - N(x_\nu,\vartheta_0,\alpha_0)] \quad (5.13)$$
$$= o_p(n^0) \quad (P_{\vartheta_0,\tau_0}^n).$$

By the same argument as in the proof of Proposition L.10 and Corollary L.11 this implies

$$m_n^{-1/2} \sum_1^{m_n} [N(x_\nu,\vartheta^{(n)}(\underline{x}),\alpha^{(n-m_n)}(x_{m_n+1},\ldots,x_n)) \quad (5.14')$$
$$- N(x_\nu,\vartheta^{(n)}(\underline{x}),\alpha_0)] = o_p(n^0) \quad (P_{\vartheta_0,\tau_0}^n).$$

Define

$$\alpha_\nu^{(n)}(\underline{x}) := \begin{cases} \alpha^{(n-m_n)}(x_{m_n+1},\ldots,x_n) & \nu \in \{1,\ldots,m_n\} \\ \alpha^{(m_n)}(x_1,\ldots,x_{m_n}) & \nu \in \{m_n+1,\ldots,n\}. \end{cases}$$

From (5.14'), together with the corresponding relation for $\nu = m_n+1,\ldots,n$, we obtain

$$n^{-1/2} \sum_1^n [N(x_\nu,\vartheta^{(n)}(\underline{x}),\alpha_\nu^{(n)}(\underline{x})) - N(x_\nu,\vartheta^{(n)}(\underline{x}),\alpha_0)] \quad (5.14)$$
$$= o_p(n^0) \quad (P_{\vartheta_0,\tau_0}^n).$$

Throughout the following we assume that $\vartheta^{(n)}(\underline{x})$ is permutation invariant. Thanks to Lemma L.17 applied with f replaced by $n^{1/2}(\vartheta^{(n)} - \vartheta_0)$, this can be done w.l.g.

For $\underline{x} \in X^n$ let $(\hat{\alpha}_1^{(n)}(\underline{x}),\ldots,\hat{\alpha}_n^{(n)}(\underline{x}))$ be chosen such that

$$\sum_1^n N(x_\nu,\vartheta^{(n)}(\underline{x}),\hat{\alpha}_\nu^{(n)}(\underline{x}))$$

is the median of $\{\sum_1^n N((\pi_n \underline{x})_\nu, \vartheta^{(n)}(\underline{x}), \alpha_\nu^{(n)}(\pi_n \underline{x})) : \pi_n \in \Pi_n\}$, where Π_n denotes the class of all permutations $\pi_n : X^n \to X^n$. According to Lemma L.17, relation (5.14) implies

$$n^{-1/2} \sum_1^n [N(x_\nu, \vartheta^{(n)}(\underline{x}), \hat{\alpha}_\nu^{(n)}(\underline{x})) - N(x_\nu, \vartheta^{(n)}(\underline{x}), \alpha_0)] \qquad (5.15)$$
$$= o_p(n^0) \qquad (P_{\vartheta_0, \tau_0}^n).$$

W.l.g. we may assume that the distance function $d|A^2$ is bounded. Since $\alpha^{(n)}(\pi_n \underline{x}) = \alpha_0 + o_p(n^0)$ $(P_{\vartheta_0, \tau_0}^n)$ for every $\pi_n \in \Pi_n$, this implies

$$\frac{1}{n!} \sum_{\pi_n} d(\alpha^{(n)}(\pi_n \underline{x}), \alpha_0) = o_p(n^0) \qquad (P_{\vartheta_0, \tau_0}^n). \qquad (5.16)$$

(Observe that $\alpha^{(n)}(\pi_n \underline{x}) - \alpha_0$ has under $P_{\vartheta_0, \tau_0}^n$ the same distribution for any $\pi_n \in \Pi_n$, n fixed.)

Let

$$\Delta^{(n)}(\underline{x}, \alpha) \qquad (5.17)$$
$$:= n^{-1/2} \Big| \sum_1^n [N(x_\nu, \vartheta^{(n)}(\underline{x}), \hat{\alpha}_\nu^{(n)}(\underline{x})) - N(x_\nu, \vartheta^{(n)}(\underline{x}), \alpha)] \Big|$$
$$+ \frac{1}{n!} \sum_{\pi_n} d(\alpha^{(n)}(\pi_n \underline{x}), \alpha).$$

By (5.15) and (5.16),

$$\Delta^{(n)}(\underline{x}, \alpha_0) = o_p(n^0) \qquad (P_{\vartheta_0, \tau_0}^n). \qquad (5.18)$$

For $\underline{x} \in X^n$ let $\hat{\alpha}^{(n)}(\underline{x})$ be such that

$$\Delta^{(n)}(\underline{x}, \hat{\alpha}^{(n)}(\underline{x})) \leq n^{-1} + \inf_{\alpha \in A} \Delta^{(n)}(\underline{x}, \alpha). \qquad (5.19)$$

Since $\Delta^{(n)}(\underline{x}, \alpha)$ is permutation invariant, $\hat{\alpha}^{(n)}(\underline{x})$ can be chosen permutation invariant, too.

It is easy to see that

$$n^{-1/2} \Big| \sum_1^n [N(x_\nu, \vartheta^{(n)}(\underline{x}), \hat{\alpha}^{(n)}(\underline{x})) - N(x_\nu, \vartheta^{(n)}(\underline{x}), \alpha_0)] \Big| \qquad (5.20)$$
$$+ d(\hat{\alpha}^{(n)}(\underline{x}), \alpha_0)$$
$$\leq \Delta^{(n)}(\underline{x}, \hat{\alpha}^{(n)}(\underline{x})) + \Delta^{(n)}(\underline{x}, \alpha_0) \leq n^{-1} + 2\Delta^{(n)}(\underline{x}, \alpha_0).$$

Because of (5.18), we have $\hat{\alpha}^{(n)} = \alpha_0 + o_p(n^0)$, and

$$n^{-1/2} \sum_1^n [N(x_\nu, \vartheta^{(n)}(\underline{x}), \hat{\alpha}^{(n)}(\underline{x})) - N(x_\nu, \vartheta^{(n)}(\underline{x}), \alpha_0)] = o_p(n^0) \quad (P^n_{\vartheta_0, \tau_0}).$$

□

Theorem 5.6 together with Proposition 5.12 imply the validity of the improvement procedure (5.7) under rather weak conditions. Though it is usually relatively easy to obtain a $\sqrt{n}$-consistent e.s. $\vartheta^{(n)}$, it might be of theoretical interest that a consistent e.s. suffices as a starting point.

Proposition 5.21. Let A_{τ_0} be a separable metric space. Assume that, in addition to (5.2), the following conditions are fulfilled: 5.12(i), (ii) and (iv), and

(i) $P_{\vartheta_0, \tau_0}(N^\bullet(\cdot, \vartheta_0, \alpha_0)) \neq 0$.

If there exists $\vartheta^{(n)} = \vartheta_0 + o_p(n^0)$ $(P^n_{\vartheta_0, \tau_0})$, then there exists a permutation invariant $\tilde{\vartheta}^{(n)} = \vartheta_0 + O_p(n^{-1/2})$ $(P^n_{\vartheta_0, \tau_0})$.

Proof. For $\delta, \vartheta \in \Theta$ let

$$\Delta^{(n)}(\underline{x}, \delta, \vartheta) := n^{1/4}|\delta - \vartheta| + m_n^{-1/2} \Big| \sum_1^{m_n} N(x_\nu, \vartheta, \alpha^{(n-m_n)}(x_{m_n+1}, \ldots, x_n)) \Big|,$$

with $m_n \in \{1, \ldots, n\}$ chosen such that $n^{-1} m_n$, $n \in \mathbb{N}$, is bounded away from 0 and 1.

Choose $\vartheta_1^{(n)}(\underline{x})$ fulfilling

$$\Delta^{(n)}(\underline{x}, \vartheta^{(n)}(\underline{x}), \vartheta_1^{(n)}(\underline{x})) \leq n^{-1} + \inf_{\vartheta \in \Theta} \Delta^{(n)}(\underline{x}, \vartheta^{(n)}(\underline{x}), \vartheta). \quad (5.22)$$

This implies, in particular,

$$\Delta^{(n)}(\underline{x}, \vartheta^{(n)}(\underline{x}), \vartheta_1^{(n)}(\underline{x})) \leq n^{-1} + \Delta^{(n)}(\underline{x}, \vartheta^{(n)}(\underline{x}), \vartheta_0). \quad (5.23)$$

In the following, all O_p and o_p are with respect to $P^n_{\vartheta_0, \tau_0}$.

Since $m_n^{-1/2} \sum_1^{m_n} N(x_\nu, \vartheta_0, \alpha_0) = O_p(n^0)$, we obtain from (5.13) that

$$m_n^{-1/2} \sum_1^{m_n} N(x_\nu, \vartheta_0, \alpha^{(n-m_n)}(x_{m_n+1}, \ldots, x_n)) = O_p(n^0). \quad (5.24)$$

Therefore, (5.23) implies

$$n^{1/4}|\vartheta^{(n)}(\underline{x}) - \vartheta_1^{(n)}(\underline{x})| \leq n^{1/4}|\vartheta^{(n)}(\underline{x}) - \vartheta_0| + O_p(n^0),$$

hence also

$$|\vartheta^{(n)}(\underline{x}) - \vartheta_1^{(n)}(\underline{x})| = o_p(n^0)$$

and

$$|\vartheta_1^{(n)}(\underline{x}) - \vartheta_0| = o_p(n^0). \tag{5.25}$$

Since

$$m_n^{-1/2} \sum_1^{m_n} N\bigl(x_\nu, \vartheta_1^{(n)}(\underline{x}), \alpha^{(n-m_n)}(x_{m_n+1}, \ldots, x_n)\bigr)$$

$$= m_n^{-1/2} \sum_1^{m_n} N\bigl(x_\nu, \vartheta_0, \alpha^{(n-m_n)}(x_{m_n+1}, \ldots, x_n)\bigr)$$

$$+ n^{1/2}\bigl(\vartheta_1^{(n)}(\underline{x}) - \vartheta_0\bigr) a_n(\underline{x})$$

with

$$a_n(\underline{x}) := (nm_n)^{-1/2} \sum_1^n \int_0^1 N^{\bullet}\bigl(x_\nu, (1-u)\vartheta_0 + u\vartheta^{(n)}(\underline{x}),$$

$$\alpha^{(n-m_n)}(x_{m_n+1}, \ldots, x_n)\bigr) du$$

we obtain

$$m_n^{-1/2} \Bigl|\sum_1^{m_n} N\bigl(x_\nu, \vartheta_1^{(n)}(\underline{x}), \alpha^{(n-m_n)}(x_{m_n+1}, \ldots, x_n)\bigr)\Bigr|$$

$$\geq -m_n^{-1/2} \Bigl|\sum_1^{m_n} N\bigl(x_\nu, \vartheta_0, \alpha^{(n-m_n)}(x_{m_n+1}, \ldots, x_n)\bigr)\Bigr|$$

$$+ n^{1/2}|(\vartheta_1^{(n)}(\underline{x}) - \vartheta_0) a_n(\underline{x})|.$$

Hence (5.23) implies

$$n^{1/2}|(\vartheta_1^{(n)}(\underline{x}) - \vartheta_0) a_n(\underline{x})| \tag{5.26}$$

$$\leq n^{-1} + n^{1/4}|\vartheta^{(n)}(\underline{x}) - \vartheta_0| + 2m_n^{-1/2}\Bigl|\sum_1^{m_n} N\bigl(x_\nu, \vartheta_0, \alpha^{(n-m_n)}(x_{m_n+1}, \ldots, x_n)\bigr)\Bigr|.$$

Because of 5.12(i) and (5.23), the sequence a_n, $n \in \mathbb{N}$, converges stochastically to $P_{\vartheta_0,\tau_0}\bigl(N^{\bullet}(\cdot, \vartheta_0, \alpha_0)\bigr)$ by Lemma L.6 and Proposition L.9 and is, therefore, bounded away from 0 stochastically. Hence (5.24) and (5.26) imply

$$|\vartheta_1^{(n)}(\underline{x}) - \vartheta_0| = o_p(n^{-1/4}). \tag{5.27}$$

Now we start the improvement procedure (5.22) once more, with $\vartheta^{(n)}(x)$ replaced by $\vartheta_1^{(n)}(x)$, thus defining an e.s. $\vartheta_2^{(n)}(x)$. We end up with (5.26), with $\vartheta^{(n)}(x)$ replaced by $\vartheta_1^{(n)}(x)$ and $\vartheta_1^{(n)}(x)$ replaced by $\vartheta_2^{(n)}(x)$.

Because of (5.24) and (5.27),
$$|\vartheta_2^{(n)}(x) - \vartheta_0| = O_p(n^{-1/2}).$$

According to Lemma L.17, applied with $f(x) = n^{1/2}(\vartheta_2^{(n)}(x) - \vartheta_0)$, $\vartheta_2^{(n)}$, $n \in \mathbb{N}$, can be replaced by a permutation invariant version. $\square$

For statisticians not abhorring artificial constructions, we formulate the following

Corollary 5.28. *Let A_{τ_0} be a separable metric space. Assume that, in addition to condition (5.2), the following conditions are fulfilled.*

(i) $(x, \vartheta, \alpha) \to N^\bullet(x, \vartheta, \alpha)$ *fulfills* $L.5((\vartheta_0, \alpha_0), P_{\vartheta_0, \tau_0})$,

(ii) $(x, \alpha) \to N(x, \vartheta_0, \alpha)^2$ *fulfills* $L.5(\alpha_0, P_{\vartheta_0, \tau_0})$,

(iii) $P_{\vartheta_0, \tau_0}(N^\bullet(\cdot, \vartheta_0, \alpha_0)) \neq 0$,

(iv) $\vartheta^{(n)} = \vartheta_0 + o_p(n^0) \quad (P_{\vartheta_0, \tau_0}^n)$,

(v) $\alpha^{(n)} = \alpha_0 + o_p(n^0) \quad (P_{\vartheta_0, \tau_0}^n)$.

Then there exists an e.s. for ϑ_0 which is as. normal with variance (5.8).

$$* \quad * \quad *$$

The results of this section cover in particular the case $A = T$, $N(\cdot, \vartheta, \tau) = L(\cdot, \vartheta, \tau)$, provided condition (5.2), i.e.
$$P_{\vartheta_0, \tau_0}(L(\cdot, \vartheta_0, \tau)) = 0 \quad \text{for all } \tau \in T,$$
is fulfilled (together with the other regularity conditions). In this case, the e.s. given by (5.7) is as. efficient, with as. variance (see (5.8)) equal to $1/P_{\vartheta_0, \tau_0}(L(\cdot, \vartheta_0, \tau_0)^2)$.

Other applications start from a general family $N(\cdot, \vartheta, \alpha)$, $\alpha \in A$. The results refer to an e.s. $\alpha^{(n)}$, $n \in \mathbb{N}$, converging to some $\alpha_0 \in A$. It suggests itself to choose an e.s. converging to a value $\alpha = \alpha(\vartheta, \tau)$ which minimizes $\alpha \to \sigma^2(\vartheta, \tau, \alpha)$ (given by (5.8)). This leads to the as. best e.s. which can be based on the family $N(\cdot, \vartheta, \alpha)$, $\alpha \in A$.

A natural way to such an e.s. is to minimize the sample–analogue of (5.8), i.e.

$$\alpha \to \frac{n^{-1} \sum_{1}^{n} N(x_\nu, \vartheta^{(n)}(\underline{x}), \alpha)^2}{\left(n^{-1} \sum_{1}^{n} N^\bullet(x_\nu, \vartheta^{(n)}(\underline{x}), \alpha)\right)^2}. \tag{5.29}$$

Conditions under which the resulting e.s. $\alpha^{(n)}$ converges to the minimizing α_0 will be given in Lemma L.33.

Remark 5.30. In general, $\sigma^2(\vartheta, \tau, \alpha(\vartheta, \tau))$ will be larger than the as. variance bound given by (4.8). In our particular case, this can also be seen directly. Since $N(\cdot, \vartheta, \alpha)$ is orthogonal to $T_0(P_{\vartheta,\tau})$, we obtain (see (4.4))

$$P_{\vartheta,\tau}(N(\cdot, \vartheta, \alpha)L(\cdot, \vartheta, \tau)) = P_{\vartheta,\tau}(N(\cdot, \vartheta, \alpha)\ell^\bullet(\cdot, \vartheta, \tau)).$$

Together with (5.4) this leads to an alternative expression for (5.8), namely

$$\sigma^2(\vartheta, \tau, \alpha) = P_{\vartheta,\tau}(N(\cdot, \vartheta, \alpha)^2)/(P_{\vartheta,\tau}(N(\cdot, \vartheta, \alpha)L(\cdot, \vartheta, \tau)))^2.$$

By Schwarz's inequality, this implies

$$\sigma^2(\vartheta, \tau, \alpha) \geq 1/P_{\vartheta,\tau}(L(\cdot, \vartheta, \tau)^2),$$

with equality iff $N(\cdot, \vartheta, \alpha)$ is proportional to $L(\cdot, \vartheta, \tau)$. This suggests to choose the family $N(\cdot, \vartheta, \alpha)$, $\alpha \in A$, large enough so that it contains functions which are close to a multiple of $L(\cdot, \vartheta, \tau)$. Without particular care in the choice of $N(\cdot, \vartheta, \alpha)$, $\alpha \in A$, there is no guarantee that the "improved" e.s. (defined by (5.7)) is as. better than the preliminary one.

The following proposition provides a characterization of the minimizing function $N(\cdot, \vartheta, \alpha(\vartheta, \tau))$. Observe that the as. variance (5.8) depends only on the direction determined by $N(\cdot, \vartheta, \alpha)$, not on its length.

Remark 5.31. If $N(\cdot, \vartheta, \alpha)$ is orthogonal to $T_0(P_{\vartheta,\tau})$ for any $\alpha \in A$, then the convex closure of $\mathcal{N} = \{N(\cdot, \vartheta, \alpha) : \alpha \in A\}$ shares this property. Hence one could always assume w.l.g. that $\mathcal{N}$ is a convex cone. This would enlarge the class of gradients, thus leading to as. better e.s. At the same time it would make it more complicated in certain cases to determine the value of α minimizing (5.29) in this larger class.

Proposition 5.32. $\alpha_0 \in A$ *mimimizes the as. variance* $\alpha \to \sigma^2(\vartheta_0, \tau_0, \alpha)$ *given by (5.8) iff* $N(\cdot, \vartheta_0, \alpha_0)$ *is proportional to the projection of* $\ell^\bullet(\cdot, \vartheta_0, \tau_0)$ *(equivalently: of* $L(\cdot, \vartheta_0, \tau_0)$*) into the cone generated by* $\{N(\cdot, \vartheta_0, \alpha) : \alpha \in A\}$.

Denoting this projection by N_0 $(= \lambda_0 N(\cdot, \vartheta_0, \alpha_0))$ we have (from (5.4))

$$P_{\vartheta_0, \tau_0}(N_0 \ell^\bullet(\cdot, \vartheta_0, \tau_0)) = P_{\vartheta_0, \tau_0}(N_0^2).$$

Hence the as. variance (5.8) may also be written as $1/P_{\vartheta_0, \tau_0}(N_0^2)$.

Proof. The proposition is an immediate consequence of the following elementary relations for a real Hilbert space H (applied for $H = L_2(P_{\vartheta_0, \tau_0})$).

Let $B \subset H$ be an arbitrary subset. Assume for some $a_0 \in H$ there exist $\lambda_0 \in \mathbb{R}$, $b_0 \in B$ such that

$$\|a_0 - \lambda_0 b_0\| \le \|a_0 - \lambda b\| \quad \text{for all } \lambda \in \mathbb{R},\ b \in B. \tag{5.33}$$

Then $\lambda_0 = (a_0, b_0)/\|b_0\|^2$ (from (5.33), applied with $b = b_0$). Applied with $\lambda = (a_0, b)/\|b\|^2$, relation (5.33) yields

$$\frac{\|b_0\|^2}{(a_0, b_0)^2} \le \frac{\|b\|^2}{(a_0, b)^2} \quad \text{for all } b \in B.$$

Hence the projection b_0 occurring in (5.33) mimimizes $b \to \frac{\|b\|^2}{(a_0,b)^2}$. The converse follows by the same type of arguments.

Finally, $(a_0, \lambda_0 b_0) = \|\lambda_0 b_0\|^2$, hence

$$\frac{\|b_0\|^2}{(a_0, b_0)^2} = \frac{1}{\|\lambda_0 b_0\|^2}.$$

To prove the assertion, apply these relations with $B = \{N(\cdot, \vartheta_0, \alpha) : \alpha \in A\}$, $a_0 = \ell^\bullet(\cdot, \vartheta_0, \tau_0)$ and $b_0 = N(\cdot, \vartheta_0, \alpha_0)$. □

Remark 5.34. If $A \subset T$ and $N(\cdot, \vartheta, \alpha) = L(\cdot, \vartheta, \tau)$, then the value $\alpha(\vartheta_0, \tau_0)$ minimizing $\alpha \to \sigma^2(\vartheta_0, \tau_0, \alpha)$ for $\alpha \in A$ is equal to τ_0, provided $\tau_0 \in A$. In this case it is irrelevant whether we use the e.s. $\alpha^{(n)}$ which minimizes (5.29), or any other e.s. converging to τ_0.

If $\tau_0 \notin A$, we are working — perhaps deliberately — with a "wrong" model. In general, wrong models usually lead to e.s. converging to a wrong parameter value. This is, of course, also true here, but it refers to the e.s. $\alpha^{(n)}$ only, not to the improved e.s. for ϑ. One should, however, bear in mind that e.s. other than those determined by (5.29) (say maximum likelihood estimators, determined under the restriction $\alpha \in A$), converge under P_{ϑ_0,τ_0} to a value different from $\alpha(\vartheta_0, \tau_0)$. This leads to e.s. for ϑ with an as. variance exceeding $\sigma^2(\vartheta_0, \tau_0, \alpha(\vartheta_0, \tau_0))$, the minimal as. variance achievable with the family $N(\cdot, \vartheta, \alpha)$, $\alpha \in A$. The mean of the limiting distribution, however, is zero.

The question raised here will be discussed in more detail in connection with the estimation of mixing distributions (see Sections 8 and 9).

6. Estimating equations

Estimating equations are an ad hoc procedure for the construction of estimators. Notwithstanding their practical usefulness and an extensive literature on this subject, there is no theory of estimating equations, making a "method" out of an "ad hoc procedure".

The usual starting point for an estimating equation for ϑ in the presence of a nuisance parameter τ is a function $N : X \times \Theta \to \mathbb{R}$, the so-called *estimating function*, with the property

$$P_{\vartheta,\tau}(N(\cdot,\vartheta)) = 0 \quad \text{for all } \tau \in T, \vartheta \in \Theta \qquad (6.1)$$

called "unbiasedness". The resulting *estimating equation* is

$$\sum_{1}^{n} N(x_\nu, \vartheta) = 0. \qquad (6.2)$$

The "information" contained in the estimating equation is

$$I(\vartheta, \tau) := \left(P_{\vartheta,\tau}(N^\bullet(\cdot,\vartheta))\right)^2 / P_{\vartheta,\tau}(N(\cdot,\vartheta)^2). \qquad (6.3)$$

Under suitable regularity conditions, $1/I(\vartheta, \tau)$ is the as. variance of the e.s. emerging as a solution of (6.2).

Conditions which guarantee as. normality with mean 0 and variance $1/I(\vartheta, \tau)$ are easy to obtain, provided one knows that this e.s. is consistent. If the solution is known to be even $\sqrt{n}$-consistent, then its as. distribution can be obtained from Theorem 5.6, specialized for $N(\cdot, \vartheta)$ not depending on α, and applied with the solution to (6.2) as preliminary (hence also improved) e.s.

Remark 6.4. Some readers might wonder where to get functions N fulfilling (6.1). First of all, any $P_{\vartheta,\tau}$-integrable function $N(\cdot, \vartheta)$ which is ancillary for $\{P_{\vartheta,\tau} : \tau \in T\}$ can be transformed into $N(\cdot,\vartheta) - P_{\vartheta,\cdot}(N(\cdot,\vartheta))$ which fulfills (6.1). But in case an ancillary function is available, one would, perhaps, be better off with m.l. estimators based on $P_{\vartheta,\cdot} * N(\cdot, \vartheta)$. Here is an alternative. If $S(\cdot, \vartheta)$ is sufficient for $\{P_{\vartheta,\tau} : \tau \in T\}$, choose $N_0(\cdot, \vartheta) \in P_{\vartheta,\cdot}^{S(\cdot,\vartheta)}(N(\cdot,\vartheta))$ (with $P_{\vartheta,\cdot}^{S(\cdot,\vartheta)}$ denoting a version (not depending on τ) of the conditional expectation, given $S(\cdot, \vartheta)$). Then $N(x, \vartheta) - N_0(S(x,\vartheta), \vartheta)$ fulfills (6.1).

In many papers estimating equations are studied as a subject in its own right, not as a technical tool for obtaining estimators. This attitude is formulated explicitly by Godambe (1976, p. 277), stating "The most common approach to the study of point estimation is to suppose that alternative estimates are given as functions of the data and to compare the properties of the corresponding random variables... A different, more general [!], and in many ways preferable [!], approach is to study not estimates but estimating functions." Even Lindsay with his attention focussed on the resulting estimators (and not on the estimating equations) considers it a "simple and elegant path to use [(6.1)] as a requirement for estimators [!] ..." (see 1985, p. 916).

This attitude manifests itself in the invention of concepts referring to estimating equations or estimating functions (like "unbiasedness", "information uniformity", etc.) which are in no clear relationship to properties of the resulting estimators. In the present writer's opinion, properties of estimating equations are of no interest unless they are related to properties of the resulting estimators which can be expressed in terms of their distributions (asymptotically, at least).

The usefulness of estimating equations is limited by the fact that, if applicable at all, the resulting e.s. will be as. inefficient in general. Even though there is no theorem asserting the existence of $\sqrt{n}$-consistent, if inefficient, e.s. under general conditions, such e.s. are usually easy to obtain by ad hoc methods in any particular case.

Let $\mathcal{N}_\vartheta := \bigcap_{\tau \in T} \mathcal{L}_*(P_{\vartheta,\tau})$, and let $\mathcal{C}_1$ denote the class of all functions $N : X \times \Theta \to \mathbb{R}$ with $N(\cdot,\vartheta) \in \mathcal{N}_\vartheta$ for $\vartheta \in \Theta$, i.e. the class of all unbiased estimating functions which are $P_{\vartheta,\tau}$-square integrable for every $\tau \in T$. (This is, roughly speaking, what Kumon and Amari (1984, p. 447) and Amari and Kumon (1988, p. 1046–1048) call the class C_1. We write $\mathcal{C}_1$ to avoid confusion.)

Since $\mathcal{N}_\vartheta \subset \bigcap_{\tau \in T} T_0^\perp(P_{\vartheta,\tau})$, $N(\cdot,\vartheta) \in \mathcal{N}_\vartheta$ for $\vartheta \in \Theta$ implies that $N(\cdot,\vartheta)$ is proportional to a gradient of the functional $\kappa(P_{\vartheta,\tau}) = \vartheta$ at $P_{\vartheta,\tau}$ for every $\tau \in T$.

Let $N_0(\cdot,\vartheta,\tau_0)$ denote the projection of $\ell^\bullet(\cdot,\vartheta,\tau_0)$ into $\mathcal{N}_\vartheta$. With τ_0 fixed, $N_0(\cdot,\vartheta,\tau_0)$ is an estimating function in $\mathcal{C}_1$, which is under P_{ϑ,τ_0} as. optimal (in the sense of leading to as. optimal e.s.). If it happens that

$$N_0(\cdot,\vartheta,\tau) = c(\vartheta,\tau) N_0(\cdot,\vartheta,\tau_0) \qquad \text{for every } \tau \in T, \tag{6.5}$$

then $N_0(\cdot,\vartheta,\tau_0)$ is as. optimal under $P_{\vartheta,\tau}$ for every $\tau \in T$, i.e. $\mathcal{C}_1$ contains an estimating function which is as. optimal, simultaneously for all $\tau \in T$. A more familiar way of writing condition (6.5) is, perhaps, the following: For every $\tau \in T$ there exists a function $h(\cdot,\vartheta,\tau) \in \mathcal{L}_*(P_{\vartheta,\tau})$, orthogonal to $\mathcal{N}_\vartheta$, and a constant $c(\vartheta,\tau)$ such that

$$N_0(\cdot,\vartheta,\tau_0) = c(\vartheta,\tau)\ell^\bullet(\cdot,\vartheta,\tau) + h(\cdot,\vartheta,\tau) \qquad \text{for every } \tau \in T. \tag{6.6}$$

Written in this way, the optimality of estimating functions fulfilling (6.5) appears as a generalization of the Theorem in Godambe and Thompson (1974, p. 569).

Even if $\mathcal{C}_1$ contains an estimating function which is as. optimal for all $\tau \in \mathrm{T}$, the pertaining as. variance will exceed the as. variance bound, except for the case that (6.5) (or the equivalent condition (6.6)) holds with $N_0(\cdot, \vartheta, \tau)$ replaced by $L(\cdot, \vartheta, \tau)$. This seems to be the essential content of Theorem 6.6 in Amari (1987a, p. 79), Theorem 4 in Amari (1987b, p. 141) and Theorem 6 in Amari and Kumon (1988, p. 1058).

Kumon and Amari (1984) introduce a smaller class of estimating functions, called C_2, which consists of all unbiased and square integrable estimating functions with the special property that their component in direction $\ell^\bullet(\cdot, \vartheta, \tau)$ is of the form $a(\vartheta)\ell^\bullet(\cdot, \vartheta, \tau)$ (with a not depending on τ). In spite of the euphonious name "information uniformity" (p. 449) this is a property (of an estimating function, not an estimator!) hard to interpret. The ultimate motivation for introducing a subclass is, however, not unfamiliar to mathematical statisticians: If there is no optimal estimator in the larger class, there is, perhaps, one in the smaller. Needless to say, the meaningfulness of such an optimum property stands and falls with the meaningfulness of the subclass. Whether C_2 always contains an as. optimal e.s. seems to be unknown. In Theorem 1, p. 451, Kumon and Amari obtain a lower bound for the as. variance of e.s. in C_2. The authors themselves raise the question whether this bound is attainable in general. Unless (6.5) holds, it will be larger than the as. variance bound. Hence, even if there are e.s. which are as. optimal in C_2, they are in general as. inefficient in the class of all regular e.s. (See, for instance, the Examples in Section 9.)

Being aware of the fact that as. optimal e.s. cannot be obtained in general as solutions to estimating equations of the type (6.1), (6.2), some authors consider estimating equations involving (an estimate of) the nuisance parameter τ. But this is the point where the concept of an estimating equation somehow dissolves. To simply take a function $N(\cdot, \vartheta, \tau)$ with $P_{\vartheta,\tau}$-expectation 0 and to solve the estimating equation $\sum_1^n N(x_\nu, \vartheta, \tau^{(n)}(\underline{x})) = 0$ (as, for instance, in Morton (1981, p. 227) or Amari (1987b, p. 145)) is a "generalization" of (6.1), (6.2) which misses an essential point. Relation (6.1) implies that the function $N(\cdot, \vartheta)$ is orthogonal to the level space $T_0(P_{\vartheta,\tau})$ and therefore proportional to a gradient, which is not the case with an arbitrary function $N(\cdot, \vartheta, \tau)$ with $P_{\vartheta,\tau}$-expectation 0.

For the mixture model involving a sufficient statistic, Lindsay (1982, 1985) evolves the idea that estimating equations based on the "conditional score function" might lead to as. efficient e.s. (See in particular (1985), p. 920, where he states (without proof) for Example 1 an expression for the as. variance bound equivalent to (E.1.3).)

7. A special semiparametric model

In this section we study models of the type described in Section 4 under the following additional assumption:

There exists a function $S : X \times \Theta \to (Y, \mathcal{B})$ such that, for every $\vartheta \in \Theta$, the function $S(\cdot, \vartheta)$ is sufficient for the family $\{P_{\vartheta,\tau} : \tau \in \mathrm{T}\}$.

We write $R_{\vartheta,\tau} := P_{\vartheta,\tau} * S(\cdot, \vartheta)$. Conditional expectations under $P_{\vartheta,\tau}$, given $S(\cdot, \vartheta)$, will be denoted by $P_{\vartheta,\cdot}^{S(\cdot,\vartheta)}$ to emphasize their independence of τ.

If such a sufficient statistic exists, $P_{\vartheta,\tau}$ admits a μ–density of the following type:
$$p(x, \vartheta, \tau) = q(x, \vartheta) p_0(S(x, \vartheta), \vartheta, \tau). \tag{7.1}$$

The representation (7.1) is not unique, for we can always rewrite it as
$$p(x, \vartheta, \tau) = \frac{q(x, \vartheta)}{C(S(x, \vartheta), \vartheta)} \cdot C(S(x, \vartheta), \vartheta) p_0(S(x, \vartheta), \vartheta, \tau).$$

Lemma 7.2. *If there exists a σ–finite $\nu | \mathcal{B}$ dominating $\mu * S(\cdot, \vartheta)$ for every $\vartheta \in \Theta$, then there exists a function $q_0 : Y \times \Theta \to [0, \infty)$ such that $q_0(\cdot, \vartheta) p_0(\cdot, \vartheta, \tau)$ is a ν–density of $R_{\vartheta,\tau}$.*

The essential point of this lemma is that q_0 does not depend on τ. As a particular consequence, we may - if convenient - always assume that p_0 in (7.1) is a ν–density of $R_{\vartheta,\tau}$.

(Rewrite (7.1) as
$$\frac{q(x, \vartheta)}{q_0(S(x, \vartheta), \vartheta)} \cdot q_0(S(x, \vartheta), \vartheta) p_0(S(x, \vartheta), \vartheta, \tau).$$

Observe that for μ–a.a. $x \in X$, $q_0(S(x, \vartheta), \vartheta) = 0$ implies $q(x, \vartheta) = 0$.)

Proof of Lemma 7.2.

To establish the existence of such a version we remark that the argument indicated in I, p. 232, carries over to the case of a sufficient statistic depending on ϑ. To see this, observe that $B \to \int 1_B(S(x, \vartheta)) q(x, \vartheta) \mu(dx)$ defines a measure on $\mathcal{B}$ which is absolutely continuous with respect to ν. By the Radon–Nikodym theorem there exists a ν–density $q_0(\cdot,\vartheta)$, say, i.e.

$$\int 1_B\bigl(S(x,\vartheta)\bigr)q(x,\vartheta)\mu(dx) = \int 1_B(s)q_0(s,\vartheta)\nu(ds), \qquad B \in \mathcal{B}.$$

It is straightforward to see that $q_0(\cdot,\vartheta)p_0(\cdot,\vartheta,\tau)$ is a ν–density of $R_{\vartheta,\tau}$. $\square$

Remark 7.3. There are important examples (see Section 9) in which the sufficient statistic $S(\cdot,\vartheta)$ can be chosen in such a way that

$$p(x,\vartheta,\tau) = q(x,\vartheta)p_0\bigl(S(x,\vartheta),\tau\bigr),$$

i.e. $p_0(\cdot,\tau)$ does not depend on ϑ. Assume there exists a measure ν dominating $\mu * S(\cdot,\vartheta)$ for $\vartheta \in \Theta$. If $\{R_{\vartheta_0,\tau} : \tau \in \mathrm{T}\}$ is complete for some $\vartheta_0 \in \Theta$, then ν can be chosen such that $p_0(\cdot,\tau)$ is a ν–density of $R_{\vartheta,\tau}$, i.e. $R_{\vartheta,\tau}$ depends on τ only.

This can be seen as follows. By Lemma 7.2, for every $\vartheta \in \Theta$ there exists $q_0(\cdot,\vartheta)$ such that $q_0(\cdot,\vartheta)p_0(\cdot,\tau)$ is a ν–density of $R_{\vartheta,\tau}$, so that

$$\int q_0(y,\vartheta)p_0(y,\tau)\nu(dy) = 1.$$

This implies

$$0 = \int \bigl(q_0(y,\vartheta) - q_0(y,\vartheta_0)\bigr)p_0(y,\tau)\nu(dy) = \int \left(\frac{q_0(y,\vartheta)}{q_0(y,\vartheta_0)} - 1\right) R_{\vartheta_0,\tau}(dy).$$

Since this holds true for all $\tau \in \mathrm{T}$, completeness implies $q_0(\cdot,\vartheta) = q_0(\cdot,\vartheta_0)$ $R_{\vartheta_0,\tau}$–a.e. Hence the assertion follows with the dominating measure $d\nu_0 = q_0(\cdot,\vartheta_0)d\nu$.

Let $\mathcal{L}_S(P_{\vartheta,\tau})$ denote the class of all functions in $\mathcal{L}_*(P_{\vartheta,\tau})$ depending on x through $S(x,\vartheta)$. (The formally consequent notation, $\bigl(\mathcal{L}_*(R_{\vartheta,\tau})\bigr) \circ S(\cdot,\vartheta)$, is somewhat clumsy.)

Let $T_*(R_{\vartheta,\tau})$ denote the tangent space of the family $\{R_{\vartheta,\hat\tau} : \hat\tau \in \mathrm{T}\}$ at $R_{\vartheta,\tau}$. Then (see (7.1)) the level space is $T_0(P_{\vartheta,\tau}) = \bigl(T_*(R_{\vartheta,\tau})\bigr) \circ S(\cdot,\vartheta)$, hence $T_0(P_{\vartheta,\tau}) \subset \mathcal{L}_S(P_{\vartheta,\tau})$. In general, this holds with $\subsetneq$. (Think of the case where τ is a finite dimensional parameter.) But there are important cases where the family $\{R_{\vartheta,\hat\tau} : \hat\tau \in \mathrm{T}\}$ is full, i.e. $T_*(R_{\vartheta,\tau}) = \mathcal{L}_*(R_{\vartheta,\tau})$, so that

$$T_0(P_{\vartheta,\tau}) = \mathcal{L}_S(P_{\vartheta,\tau}). \tag{7.4}$$

In Section 8 this will be shown for mixture models. Van der Vaart (1988, Theorem 5.10) establishes (7.4) by adding to completeness the condition that the family $\{P_{\vartheta,\tau} :$

$\tau \in T\}$ is closed under convex combinations, and that $p_0(\cdot, \vartheta, \tau)/p_0(\cdot, \vartheta, \tau_0)$ is R_{ϑ, τ_0}-square integrable for any $\tau \in T$. Under these assumptions, one can consider the path $\tau_t \to \tau_0$, defined by

$$p_0(\cdot, \vartheta, \tau_t) = (1-t)p_0(\cdot, \vartheta, \tau_0) + tp_0(\cdot, \vartheta, \tau),$$

which has derivative $\frac{p_0(\cdot, \vartheta, \tau)}{p_0(\cdot, \vartheta, \tau_0)} - 1$. Hence $T_0(P_{\vartheta, \tau_0})$ contains all functions $\frac{p_0(\cdot, \vartheta, \tau)}{p_0(\cdot, \vartheta, \tau_0)} - 1$, $\tau \in T$. Completeness implies that the family of these functions is full.

Throughout the following we presume that $Y \subset \mathbb{R}$. The extension to $Y \subset \mathbb{R}^m$ is straightforward. (In the examples of Section 9, the cases $Y = \mathbb{R}$ and $Y = \mathbb{R}_+$ occur.)

From (7.1),

$$\ell^\bullet(x, \vartheta, \tau) = \frac{q^\bullet(x, \vartheta)}{q(x, \vartheta)} \qquad (7.5)$$
$$+ S^\bullet(x, \vartheta)\frac{p_0'(S(x,\vartheta), \vartheta, \tau)}{p_0(S(x,\vartheta), \vartheta, \tau)} + \frac{p_0^\bullet(S(x,\vartheta), \vartheta, \tau)}{p_0(S(x,\vartheta), \vartheta, \tau)}.$$

Assume that the functions $q^\bullet(\cdot, \vartheta)/q(\cdot, \vartheta)$ and $S^\bullet(\cdot, \vartheta)$ are $P_{\vartheta, \tau}$-square integrable, for every $\tau \in T$. With

$$a(\cdot, \vartheta) := q^\bullet(\cdot, \vartheta)/q(\cdot, \vartheta) - P_{\vartheta, \cdot}^{S(\cdot, \vartheta)}\big(q^\bullet(\cdot, \vartheta)/q(\cdot, \vartheta)\big) \qquad (7.6')$$

and

$$b(\cdot, \vartheta) := S^\bullet(\cdot, \vartheta) - P_{\vartheta, \cdot}^{S(\cdot, \vartheta)}\big(S^\bullet(\cdot, \vartheta)\big) \qquad (7.6'')$$

let

$$L_0(x, \vartheta, \tau) := a(x, \vartheta) + b(x, \vartheta)\frac{p_0'(S(x,\vartheta), \vartheta, \tau)}{p_0(S(x,\vartheta), \vartheta, \tau)}. \qquad (7.7)$$

Since $\ell^\bullet(\cdot, \vartheta, \tau)$ is independent of the special choice of the factors q and p_0 in the representation (7.1), the same holds true for $L_0(\cdot, \vartheta, \tau)$.

Since the conditional expectation of any square integrable function is its projection into $\mathcal{L}_S(P_{\vartheta, \tau})$, the function $L_0(\cdot, \vartheta, \tau)$ is the orthogonal component of $\ell^\bullet(\cdot, \vartheta, \tau)$ with respect to $\mathcal{L}_S(P_{\vartheta, \tau})$. For the purpose of estimating $L_0(\cdot, \vartheta, \tau)$, it is important that the following stronger condition holds (which is an immediate consequence of (7.6))

$$P_{\vartheta, \cdot}^{S(\cdot, \vartheta)}\big(L_0(\cdot, \vartheta, \hat{\tau})\big) = 0 \qquad \text{for all } \hat{\tau} \in T. \qquad (7.8)$$

Hence $L_0(\cdot, \vartheta, \hat{\tau})$ is orthogonal to $\mathcal{L}_S(P_{\vartheta, \tau})$ for any $\hat{\tau} \in T$. Since $T_0(P_{\vartheta, \tau}) \subset \mathcal{L}_S(P_{\vartheta, \tau})$, the function $L_0(\cdot, \vartheta, \hat{\tau})$ is, in particular, orthogonal to $T_0(P_{\vartheta, \tau})$ for any $\hat{\tau} \in T$. Hence the results of Section 5 based on the family $\{N(\cdot, \vartheta, \alpha) : \alpha \in A\}$

apply for $A = T$ and $N(\cdot, \vartheta, \tau) := L_0(\cdot, \vartheta, \tau)$, provided this function fulfills the pertinent regularity conditions.

The resulting e.s. will be as. normal with as. variance

$$\sigma^2(\vartheta, \tau) = 1/P_{\vartheta,\tau}\big(L_0(\cdot, \vartheta, \tau)^2\big). \tag{7.9}$$

(Since $L_0(\cdot, \vartheta, \tau)$ is the projection of $\ell^\bullet(\cdot, \vartheta, \tau)$ into the cone generated by $\{L_0(\cdot, \vartheta, \hat{\tau}): \hat{\tau} \in T\}$, this follows from Proposition 5.32.)

If $T_0(P_{\vartheta,\tau}) = \mathcal{L}_S(P_{\vartheta,\tau})$, then $L_0(\cdot, \vartheta, \tau)$ is the orthogonal component of $\ell^\bullet(\cdot, \vartheta, \tau)$ with respect to $T_0(P_{\vartheta,\tau})$, hence proportional to the *canonical* gradient. Then the resulting e.s. are as. efficient, and the as. variance given by (7.9) equals the as. variance bound.

Remark 7.10. In this connection we ought to mention a "degenerate" case which occurs if $S(\cdot, \vartheta)$ does not depend on ϑ. If we standardize representation (7.1) to have $p_0(\cdot, \vartheta, \tau)$ a ν–density of $R_{\vartheta,\tau}$ (see the remark following Lemma 7.2), then

$$P_{\vartheta,\tau}\big(q(\cdot, \delta)/q(\cdot, \vartheta)\big) = 1 \quad \text{for all } \delta \in \Theta. \tag{7.11}$$

This can be seen as follows:

$$\begin{aligned} 1 &= \nu\big(p_0(\cdot, \vartheta, \tau)\big) = R_{\delta,\tau}\big(p_0(\cdot, \vartheta, \tau)/p_0(\cdot, \delta, \tau)\big) \\ &= P_{\delta,\tau}\big(p_0(S(\cdot), \vartheta, \tau)/p_0(S(\cdot), \delta, \tau)\big) \\ &= \mu\big(q(\cdot, \delta)p_0(S(\cdot), \vartheta, \tau)\big) = P_{\vartheta,\tau}\big(q(\cdot, \delta)/q(\cdot, \vartheta)\big). \end{aligned}$$

Since the function 'log' is strictly concave, we obtain from (7.11) by Jensen's inequality that

$$P_{\vartheta,\tau}\big(\log q(\cdot, \delta)\big) < P_{\vartheta,\tau}\big(\log q(\cdot, \vartheta)\big) \quad \text{for all } \tau \in T, \ \delta \neq \vartheta.$$

Under suitable regularity conditions, the "partial" maximum likelihood estimator, obtained by maximization of $\delta \to \sum_1^n \log q(x_\nu, \delta)$, is as. normal with variance $1/P_{\vartheta,\tau}\big((q^\bullet(\cdot, \vartheta)/q(\cdot, \vartheta))^2\big)$.

If (7.11) may be differentiated with respect to δ under the integral, we obtain $P_{\vartheta,\tau}\big(q^\bullet(\cdot, \vartheta)/q(\cdot, \vartheta)\big) = 0$. Since this relation holds for all $\tau \in T$, we obtain under suitable differentiability conditions for paths $\tau_t \to \tau$ that $q^\bullet(\cdot, \vartheta)/q(\cdot, \vartheta)$ is orthogonal to $T_0(P_{\vartheta,\tau})$.

Since

$$\ell^\bullet(x, \vartheta, \tau) = \frac{q^\bullet(x, \vartheta)}{q(x, \vartheta)} + \frac{p_0^\bullet\big(S(x), \vartheta, \tau\big)}{p_0\big(S(x), \vartheta, \tau\big)},$$

$q^\bullet(\cdot, \vartheta)/q(\cdot, \vartheta)$ is the orthogonal component of $\ell^\bullet(\cdot, \vartheta, \tau)$ with respect to $T_0(P_{\vartheta,\tau})$ iff $p_0^\bullet\big(S(\cdot), \vartheta, \tau\big)/p_0\big(S(\cdot), \vartheta, \tau\big) \in T_0(P_{\vartheta,\tau})$.

Hence the condition $p_0^\bullet(S(\cdot),\vartheta,\tau)/p_0(S(\cdot),\vartheta,\tau) \in T_0(P_{\vartheta,\tau})$ is necessary for the partial maximum likelihood estimator to be as. efficient (and sufficient if regularity conditions for $q^\bullet/q$ are fulfilled which guarantee that the partial maximum likelihood estimator is as. normal with variance $1/P_{\vartheta,\tau}((q^\bullet(\cdot,\vartheta)/q(\cdot,\vartheta))^2)$). With a finite dimensional parameter τ such a condition will be fulfilled in exceptional cases only. (See Andersen (1970, p. 297) or Liang (1983, Section 4).) $p_0^\bullet(S(\cdot),\vartheta,\tau)/p_0(S(\cdot),\vartheta,\tau) \in T_0(P_{\vartheta,\tau})$ is a matter of course if condition (7.4) holds.

See also I 14.3 for a more detailed discussion for the case of mixture models.

*   *

*

Now we return to the general case with $S(\cdot,\vartheta)$ depending on ϑ.

The *practical* way is to choose a sufficiently large subfamily of T and to estimate the value of τ in this subfamily for which the as. variance becomes minimal (see Remark 5.34 and Section 8).

Of rather more *theoretical* interest is the fact that (under suitable regularity conditions on L_0 given in Theorem 5.6 and Proposition 5.12, and in the presence of a $\sqrt{n}$-consistent preliminary e.s. for ϑ) a *consistent* e.s. for τ is all we need to construct e.s. for ϑ with as. variance (7.9). The serious drawback of Proposition 5.12 is the use of the splitting trick, and the question naturally arises whether one can do without this under the more special conditions of the model under consideration. This is, in fact, the case if one is ready to exchange the splitting trick for the discretization trick (invented by LeCam, 1960, Appendix 1).

Recall that the splitting trick in the proof of Proposition 5.12 is needed to overcome the difficulties resulting from the dependence between x_ν and $\alpha^{(n)}(x_1,\ldots,x_n)$. In the presence of a sufficient statistic, one could try the following idea. If ϑ is known, τ can be estimated from $(S(x_1,\vartheta),\ldots,S(x_n,\vartheta))$, say by $\tau^{(n)}(S(x_1,\vartheta),\ldots,S(x_n,\vartheta))$. Since

$$L_0(x_\nu,\vartheta,\tau^{(n)}(S(x_1,\vartheta),\ldots,S(x_n,\vartheta)))$$

and

$$L_0(x_\mu,\vartheta,\tau^{(n)}(S(x_1,\vartheta),\ldots,S(x_n,\vartheta)))$$

are conditionally independent, given $(S(x_1,\vartheta),\ldots,S(x_n,\vartheta))$, relations (7.7), (7.8), together with condition L.5$((\vartheta,\tau),P_{\vartheta,\tau})$ for L_0 can be used in a straightforward way to show that

$$n^{-1/2}\sum_1^n \Big(L_0(x_\nu,\vartheta,\tau^{(n)}(S(x_1,\vartheta),\ldots,S(x_n,\vartheta))) - L_0(x_\nu,\vartheta,\tau)\Big) \quad (7.12)$$
$$= o_p(n^0) \quad (P_{\vartheta,\tau}^n).$$

With ϑ unknown, we need (7.12) with ϑ replaced by a preliminary e.s. $\vartheta^{(n)}$, and this destroys the "conditional independence"–argument.

The following Lemma 7.16 shows that the dependence–problem can be solved if we replace $\vartheta^{(n)}(\underline{x})$ by a discretized version. It refers to a more general family of functions in $T_0^\perp(P_{\vartheta,\tau})$, namely

$$N(x,\vartheta,\alpha) = a(x,\vartheta) + b(x,\vartheta)H\big(S(x,\vartheta),\vartheta,\alpha\big), \qquad \alpha \in A_\tau, \qquad (7.13)$$

with a, b defined by (7.6), and with some arbitrary function $H(\cdot,\vartheta,\alpha)$ for which $\big(b(\cdot,\vartheta)H(S(\cdot,\vartheta),\vartheta,\alpha)\big)^2$ is $P_{\vartheta,\tau}$–integrable for $\alpha \in A_\tau$. Throughout the following we assume that, for any $\tau \in T$, A_τ is a Hausdorff space with countable base.

For $n \in \mathbb{N}$ and $\nu \in \{1,\ldots,n\}$ let $H_\nu^{(n)} : \Theta \times Y^n \to \mathbb{R}$.

For any estimator $\vartheta^{(n)} : X^n \to \mathbb{R}$ we define the *discretized version* by

$$\vartheta_*^{(n)}(\underline{x}) := n^{-1/2}[n^{1/2}\vartheta^{(n)}(\underline{x})], \qquad \underline{x} \in X^n. \qquad (7.14)$$

Finally, we write

$$B(\cdot,\vartheta) := P_{\vartheta,\cdot}^{S(\cdot,\vartheta)}\big(b(\cdot,\vartheta)^2\big). \qquad (7.15)$$

Lemma 7.16. *Assume the following conditions.*

For any (nonstochastic) sequence $\vartheta_n = \vartheta_0 + O(n^{-1/2})$,

(i) *the sequences P_{ϑ_n,τ_0}^n, $n \in \mathbb{N}$, and P_{ϑ_0,τ_0}^n, $n \in \mathbb{N}$, are contiguous,*

(ii) $n^{-1} \sum_1^n B(s_\nu, \vartheta_n)\big(H_\nu^{(n)}(\vartheta_n; s_1,\ldots,s_n) - H(s_\nu,\vartheta_n,\alpha_0)\big)^2$
$= o_p(n^0)$ $(R_{\vartheta_n,\tau_0}^n)$ *for some $\alpha_0 \in A_{\tau_0}$.*

(iii) $\vartheta^{(n)} = \vartheta_0 + O_p(n^{-1/2})$ $(P_{\vartheta_0,\tau_0}^n)$.

Then

$$n^{-1/2} \sum_1^n b(x_\nu, \vartheta_*^{(n)}(\underline{x}))\big[H_\nu^{(n)}\big(\vartheta_*^{(n)}(\underline{x}); S(x_1,\vartheta_*^{(n)}(\underline{x})),\ldots,S(x_n,\vartheta_*^{(n)}(\underline{x}))\big)$$
$$- H\big(S(x_\nu,\vartheta_*^{(n)}(\underline{x})),\vartheta_*^{(n)}(\underline{x}),\alpha_0\big)\big] = o_p(n^0) \qquad (P_{\vartheta_0,\tau_0}^n),$$

which corresponds to Condition 5.6(iv).

Proof. Let

$$\Delta_\nu^{(n)}(\vartheta; s_1,\ldots,s_n) := H_\nu^{(n)}(\vartheta; s_1,\ldots,s_n) - H(s_\nu,\vartheta,\alpha_0)$$

and

$$A_n(\delta) :=$$
$$\{\underline{x} \in X^n : |n^{-1/2} \sum_1^n b(x_\nu, \vartheta_n) \Delta_\nu^{(n)}(\vartheta_n; S(x_1, \vartheta_n), \ldots, S(x_n, \vartheta_n))| > \delta\}.$$

We shall prove that

$$P^n_{\vartheta_n, \tau_0}(A_n(\delta)) = o(n^0) \quad \text{for every } \delta > 0. \tag{7.17}$$

This means that

$$n^{-1/2} \sum_1^n b(x_\nu, \vartheta_n) \Big(H_\nu^{(n)}(\vartheta_n; S(x_1, \vartheta_n), \ldots, S(x_n, \vartheta_n))$$
$$- H(S(x_\nu, \vartheta_n), \vartheta_n, \alpha_0) \Big) = o_p(n^0) \quad (P^n_{\vartheta_n, \tau_0}).$$

By the Discretization–Lemma L.20 and condition (i), the sequence ϑ_n may be replaced by $\vartheta_*^{(n)}(\underline{x})$ in this relation. From this, the assertion follows easily.

To prove (7.17), we proceed as follows. Since $P^{S(\cdot, \vartheta_n)}_{\vartheta_n, \cdot}(b(\cdot, \vartheta_n)) = 0$, we obtain that the conditional P_{ϑ_n, τ_0}-expectation, given $S(x_\nu, \vartheta_n)$, of

$$b(x_\nu, \vartheta_n) b(x_\mu, \vartheta_n) \Delta_\nu^{(n)}(\vartheta_n; S(x_1, \vartheta_n), \ldots, S(x_n, \vartheta_n))$$
$$\cdot \Delta_\mu^{(n)}(\vartheta_n; S(x_1, \vartheta_n), \ldots, S(x_n, \vartheta_n))$$

is 0 for $\mu \neq \nu$.

For all $s_\nu \in Y$, $\nu = 1, \ldots, n$, the conditional expectation of $1_{A_n(\delta)}$, given $S(x_\nu, \vartheta_n) = s_\nu$, $\nu = 1, \ldots, n$, is (by Čebyšev's inequality) bounded by

$$\delta^{-2} n^{-1} \sum_1^n B(s_\nu, \vartheta_n) \big(H_\nu^{(n)}(\vartheta_n; s_1, \ldots, s_n) - H(s_\nu, \vartheta_n, \alpha_0) \big)^2.$$

By 7.16(ii), this bound is $o_p(n^0)$ $(R^n_{\vartheta_n, \tau_0})$. Since the conditional expectation of $1_{A_n(\delta)}$ is bounded, (7.17) follows by $R^n_{\vartheta_n, \tau_0}$-integration. □

We now consider estimators $\alpha^{(n)}$ for α which depend on $(x_1, \ldots, x_n)$ through $S(x_1, \vartheta), \ldots, S(x_n, \vartheta)$ only. This is convenient for technical reasons, and sufficient for the purpose of estimating $H(\cdot, \vartheta, \alpha)$. The following Proposition gives conditions under which $H(\cdot, \vartheta, \alpha^{(n)})$ can be used as an estimator for $H(\cdot, \vartheta, \alpha)$.

Proposition 7.18. Assume the following conditions for any (nonstochastic) sequence $\vartheta_n = \vartheta_0 + O(n^{-1/2})$.

(i) $\hat{\alpha}^{(n)} : Y^n \to A$ fulfills $\hat{\alpha}^{(n)} = \alpha_0 + o_p(n^0)$ $(R^n_{\vartheta_n,\tau_0})$.

(ii) For every $\varepsilon > 0$ there exists a neighborhood V_ε of α_0 such that for all sufficiently large $n \in \mathbb{N}$,

$$\int B(s,\vartheta_n) \sup_{\alpha \in V_\varepsilon} \big(H(s,\vartheta_n,\alpha) - H(s,\vartheta_n,\alpha_0)\big)^2 R_{\vartheta_n,\tau_0}(ds) < \varepsilon.$$

Then condition 7.16(ii) holds with

$$H_\nu^{(n)}(\vartheta;s_1,\ldots,s_n) := H\big(s_\nu,\vartheta,\hat{\alpha}^{(n)}(s_1,\ldots,s_n)\big).$$

Proof. For any $\delta > 0$,

$$R^n_{\vartheta_n,\tau_0}\{\underline{s} \in Y^n : n^{-1}\sum_1^n B(s_\nu,\vartheta_n)\big(H(s_\nu,\vartheta_n,\hat{\alpha}^{(n)}(\underline{s})) - H(s_\nu,\vartheta_n,\alpha_0)\big)^2 > \delta\}$$

$$\leq R^n_{\vartheta_n,\tau_0}\{\underline{s} \in Y^n : \hat{\alpha}^{(n)}(\underline{s}) \notin V_\varepsilon\} + R^n_{\vartheta_n,\tau_0}\{\underline{s} \in Y^n :$$

$$n^{-1}\sum_1^n B(s_\nu,\vartheta_n) \sup_{\alpha \in V_\varepsilon}\big(H(s_\nu,\vartheta_n,\alpha) - H(s_\nu,\vartheta_n,\alpha_0)\big)^2 > \delta\}$$

$$\leq R^n_{\vartheta_n,\tau_0}\{\underline{s} \in Y^n : \hat{\alpha}^{(n)}(\underline{s}) \notin V_\varepsilon\} + \varepsilon/\delta.$$

Because of (i), the assertion follows immediately. □

Remark 7.19. Theorem 5.6 uses under 5.6(iii) an e.s. $d^{(n)}$ for

$$P_{\vartheta_0,\tau_0}(N^\bullet(\cdot,\vartheta_0,\alpha_0)).$$

Since a sufficient statistic is available, one might choose

$$d^{(n)}(\underline{x}) = n^{-1} \sum_1^n N_0\big(S(x_\nu,\vartheta^{(n)}(\underline{x})),\vartheta^{(n)}(\underline{x}),\alpha^{(n)}(\underline{x})\big),$$

with $\alpha^{(n)}(\underline{x}) := \hat{\alpha}^{(n)}\big(S(x_1,\vartheta^{(n)}(\underline{x})),\ldots,S(x_n,\vartheta^{(n)}(\underline{x}))\big)$ and $N_0(\cdot,\vartheta,\alpha) = P^{S(\cdot,\vartheta)}_{\vartheta,\cdot}(N^\bullet(\cdot,\vartheta,\alpha))$.

The e.s. for $P_{\vartheta_0,\tau_0}(N^\bullet(\cdot,\vartheta_0,\alpha_0))$ based on $P^{S(\cdot,\vartheta)}_{\vartheta,\cdot}(N^\bullet(\cdot,\vartheta,\alpha))$ has, in general, a smaller as. variance than the e.s. based on $N^\bullet(\cdot,\vartheta,\alpha)$, but the difference is as. so small that its influence on the as. distribution of the improved e.s. for ϑ evades an as. analysis of 1st order.

There is a totally different way of estimating $L_0(\cdot, \vartheta, \tau)$, which starts from the observation that τ occurs in $L_0(\cdot, \vartheta, \tau)$ only through $p'_0(\cdot, \vartheta, \tau)/p_0(\cdot, \vartheta, \tau)$. Under appropriate standardization (see Lemma 7.2), $p_0(\cdot, \vartheta, \tau)$ is a density of $R_{\vartheta, \tau}$. If ϑ were known, one could use the "observations" $S(x_1, \vartheta), \ldots, S(x_n, \vartheta)$ to estimate the density $p_0(\cdot, \vartheta, \tau)$ and its derivative $p'_0(\cdot, \vartheta, \tau)$. With ϑ unknown, one could try to apply this estimation procedure to

$$S(x_1, \vartheta^{(n)}(\underline{x})), \ldots, S(x_n, \vartheta^{(n)}(\underline{x})).$$

At first view this appears to be the ideal solution, since no further informations are required about how $p_0(\cdot, \vartheta, \tau)$ depends on τ. A natural choice in this case are kernel estimators, because of their simple structure: Since they are sums of independent functions, it is (not easy, but) feasible to establish the properties required for an estimator of the canonical gradient to be suitable for the improvement procedure.

The interest in estimating the logarithmic derivative of a density on $\mathbb{R}$ has a long tradition. It originated in a totally different context, that of "empirical Bayes" procedures. For exponential families, say $p(x, \eta) = c(\eta) \exp[\eta x]$, with prior distribution Γ, the Bayes estimate for η, given x, is

$$\int \eta p(x, \eta) \Gamma(d\eta)/p(x, \Gamma) = p'(x, \Gamma)/p(x, \Gamma)$$

(with $p(x, \Gamma) = \int p(x, \eta) \Gamma(d\eta)$).

Starting with Robbins (1963), numerous papers have been devoted to the problem of estimating $p'(x, \Gamma)/p(x, \Gamma)$. Kernel estimators with almost optimal rates can be found in Singh (see (1979) and the papers cited there).

Bickel (1982), Bickel and Ritov (1987) were the first to show (in connection with special models) that kernel estimators can be successfully applied for the estimation of $p'(\cdot, \vartheta, \Gamma)/p(\cdot, \vartheta, \Gamma)$ in connection with the improvement procedure. Van der Vaart (1988, Sections 5.2, 5.3) obtains a result for general models of type (7.1). That these results were developed independently of what was available on kernel estimators in literature is justified by the fact that estimators for p'/p play an ancillary role in the improvement procedure, and for this no rates are needed. Moreover, the function to be estimated for the improvement procedure is $b(\cdot, \vartheta) p'_0(S(\cdot, \vartheta), \vartheta, \tau)/p_0(S(\cdot, \vartheta), \vartheta, \tau)$, not $p'_0(\cdot, \vartheta, \tau)/p_0(\cdot, \vartheta, \tau)$. Finally, $p'_0(\cdot, \vartheta, \tau)/p_0(\cdot, \vartheta, \tau)$ involves the unknown parameter ϑ, which is absent in the literature on empirical Bayes procedures.

If kernel estimators are a useful tool for obtaining basic results on the existence of e.s., they appear to be less suitable for practical purposes. Theorems using kernel estimators for the improvement procedure use a certain rate at which the window width is to converge to zero. Moreover, some modifications are called for in order to prevent the estimators for p'_0/p_0 from attaining too large values. This can either be achieved by truncation of the quotient as in

Singh (1979, p. 894, 3.10) or by bounding the denominator away from 0 as in Bickel (1962, p. 665), Bickel and Ritov (1987, p. 522) or van der Vaart (1988, Section 5.3). The e.s. which finally come out of such modifications are highly artificial constructs, tailored for the proof of a mathematical theorem (asserting the as. efficiency of the improved e.s. for ϑ). In particular, they require the use of discretized preliminary e.s. for ϑ. These constructions involve several sequences of constants (3 in the paper of Bickel and Ritov, 7 in van der Vaart's paper, plus 1 for the discretization) which have to converge to zero or infinity at certain carefully balanced rates.

In applications one is interested in prescriptions for the computation of estimators, and in approximative informations about their accuracy. Constructions involving unspecified sequences of constants do not lead to such prescriptions, and the limiting distributions of highly complicated e.s. may be misleading if we interpret them as approximations to their true distribution, even for not–so–small samples. (Simulation experiments performed in connection with mixture models show that for sample sizes of several hundred the use of kernel estimators does not lead to estimators for ϑ the distribution of which is in approximate agreement with the limiting distribution.)

What appears as a striking advantage of kernel estimators — that one needs no assumptions on how $p_0'(\cdot, \vartheta, \tau)/p_0(\cdot, \vartheta, \tau)$ depends on τ — may in fact be the cause of the poor numerical results obtained in this way: Kernel estimators dispense with any information available about the function $\tau \to p_0'(\cdot, \vartheta, \tau)/p_0(\cdot, \vartheta, \tau)$ (say continuity, at least) and are for that very reason not the best choice if one is interested in applications. To take an extreme example: If τ were a real parameter, nobody would seriously consider kernel estimators for $p_0'(\cdot, \vartheta, \tau)/p_0(\cdot, \vartheta, \tau)$ despite the fact that their suitability for the improvement procedure is the same as that of $p_0'(\cdot, \vartheta, \tau^{(n)}(\underline{x}))/p_0(\cdot, \vartheta, \tau^{(n)}(\underline{x}))$ if judged by the limiting distribution of the resulting e.s. for ϑ. Even though this difference escapes an as. analysis of 1st order (i.e. the limiting distributions are the same), it will be noticeable if the true distributions (for moderate n) are compared in a simulation experiment.

8. Mixture models

In this section we turn to the special type of models (7.1) where τ is a mixing p–measure. To emphasize this specialization, we change our notation from τ to Γ.

Let $\Theta \subset \mathbb{R}$ and $(H, \mathcal{C})$ be a measurable space. For $\vartheta \in \Theta$, $\eta \in H$ let $P_{\vartheta,\eta}$ be a p–measure with μ–density $p(\cdot, \vartheta, \eta)$. For any p–measure $\Gamma | \mathcal{C}$ let $P_{\vartheta,\Gamma}$ denote the Γ–mixture of the family $P_{\vartheta,\eta}$, $\eta \in H$, defined by $P_{\vartheta,\Gamma}(A) := \int P_{\vartheta,\eta}(A) \Gamma(d\eta)$.

If $(x, \eta) \to p(x, \vartheta, \eta)$ is measurable, $p(x, \vartheta, \Gamma) := \int p(x, \vartheta, \eta) \Gamma(d\eta)$ defines a μ–density of $P_{\vartheta,\Gamma}$.

Throughout this section we assume a factorization corresponding to (7.1),

$$p(x, \vartheta, \eta) = q(x, \vartheta) p_0(S(x, \vartheta), \vartheta, \eta). \tag{8.1}$$

This implies the factorization

$$p(x, \vartheta, \Gamma) = q(x, \vartheta) p_0(S(x, \vartheta), \vartheta, \Gamma), \tag{8.1'}$$

with $p_0(s, \vartheta, \Gamma) := \int p_0(s, \vartheta, \eta) \Gamma(d\eta)$.

Hence $S(\cdot, \vartheta)$ is sufficient for $\{P_{\vartheta,\eta} : \eta \in H\}$ and therefore for $\{P_{\vartheta,\Gamma} : \Gamma \in \mathcal{G}\}$, with $\mathcal{G}$ denoting a family of p–measures over $(H, \mathcal{C})$. As in Section 7, we use $R_{\vartheta,\eta} = P_{\vartheta,\eta} * S(\cdot, \vartheta)$, and, correspondingly, $R_{\vartheta,\Gamma} = P_{\vartheta,\Gamma} * S(\cdot, \vartheta)$.

In the following we discuss those features which distinguish the mixture model (8.1') from the more general model (7.1).

First of all, we obtain an explicit expression for the level space (see Section 4), namely

$$T_0(P_{\vartheta,\Gamma}) = \left\{ x \to \frac{\int k(\eta) p_0(S(x, \vartheta), \vartheta, \eta) \Gamma(d\eta)}{p_0(S(x, \vartheta), \vartheta, \Gamma)} \ : \ k \in T(\Gamma, \mathcal{G}) \right\}. \tag{8.2}$$

In natural examples, $\mathcal{G}$ is the family of all p–measures Γ which are equivalent to some σ–finite measure γ over $(H, \mathcal{C})$. Then $T(\Gamma, \mathcal{G}) = \mathcal{L}_*(\Gamma)$.

If the family $\{R_{\vartheta,\eta} : \eta \in H\}$ is complete in the sense that for all $f \in \mathcal{L}_2(R_{\vartheta,\Gamma})$

$$R_{\vartheta,\eta}(f) = 0 \text{ for } \gamma\text{-a.a. } \eta \in H \text{ implies } f = 0 \quad R_{\vartheta,\eta}\text{-a.e.}, \tag{8.3}$$

then $T(R_{\vartheta,\Gamma}, \{R_{\vartheta,\hat{\Gamma}} : \hat{\Gamma} \in \mathcal{G}\}) = \mathcal{L}_*(R_{\vartheta,\Gamma})$, and therefore (7.4) holds, i.e. $T_0(P_{\vartheta,\Gamma}) = \mathcal{L}_S(P_{\vartheta,\Gamma})$.

(If $f \in \mathcal{L}_*(R_{\vartheta,\Gamma})$ is orthogonal to $y \to \int k(\eta)p_0(y,\vartheta,\eta)\Gamma(d\eta)/p_0(y,\vartheta,\Gamma)$, this implies $\int R_{\vartheta,\eta}(f)k(\eta)\Gamma(d\eta) = 0$. If this holds for all $k \in \mathcal{L}_*(\Gamma)$, $R_{\vartheta,\eta}(f) = 0$ Γ–a.e. follows. By the completeness condition (8.3) this implies $f = 0$ $R_{\vartheta,\eta}$–a.e.)

Next we discuss the problem of estimating $p_0'(\cdot,\vartheta,\Gamma)/p_0(\cdot,\vartheta,\Gamma)$ (which is required for the improvement procedure outlined in Theorem 5.6, Propositions 5.12 and 7.18).

The kernel estimators for $p_0(\cdot,\vartheta,\tau)$ and $p_0'(\cdot,\vartheta,\tau)$ mentioned in Section 7 make no assumptions about the nature of the parameter τ. Hence they apply, in particular, for $\tau = \Gamma$. In estimating $p_0(\cdot,\vartheta,\Gamma)$ and $p_0'(\cdot,\vartheta,\Gamma)$ directly, we miss, however, the smoothing effect involved in the integration of $\eta \to p(s,\vartheta,\eta)$ with respect to Γ. Hence it appears advisable to make use of the special character of the mixing model, i.e. to estimate the mixing distribution Γ by $\Gamma^{(n)}(\underline{x})$, and to estimate $\int p(s,\vartheta,\eta)\Gamma(d\eta)$ and $\int p'(s,\vartheta,\eta)\Gamma(d\eta)$ by

$$\int p\bigl(s,\vartheta^{(n)}(\underline{x}),\eta\bigr)\Gamma^{(n)}(\underline{x})(d\eta) \quad \text{and} \quad \int p'\bigl(s,\vartheta^{(n)}(\underline{x}),\eta\bigr)\Gamma^{(n)}(\underline{x})(d\eta),$$

respectively.

The resulting smoothing effect does not show up in an asymptotic analysis of 1st order, but it becomes manifest in simulation experiments. The results of the improvement procedure given by (5.7), if applied with kernel estimators for p_0 and p_0', are discouraging even for sample sizes of 100, say. For this reason, the use of empirical Bayes estimators for p_0 and p_0' suggests itself.

Starting from Robbins (1964), several authors have obtained e.s. $\Gamma^{(n)}$ which converge weakly to the true mixing distribution. A survey of the recent literature is Titterington (1989).

Among the papers achieving this goal by a method akin to the minimum distance method are Choi (1969), Deely and Kruse (1968), Blum and Susarla (1977), Hall (1981), etc. These results are not immediately applicable in our special context since they do not involve the additional parameter ϑ.

Another possibility is to use the m.l. (= maximum likelihood) method. Though m.l. e.s. may be useless in nonparametric models, this is not the case here, thanks to a special feature of mixture models: The map $\Gamma \to p_0(s,\vartheta,\Gamma)$ is linear. This property makes it easy to show that any sequence of as. m.l. estimators is consistent with respect to weak convergence (see Pfanzagl 1988, 1989). Moreover, the E.M. algorithm (see Little and Rubin, 1987, for a readable presentation) is an efficient tool for approximating the m.l. estimators numerically.

According to Lindsay (1983, p. 89, Theorem 3.1) the exact m.l. estimator for Γ, based on a sample of size n, is a discrete distribution with at most n

supporting points. If $R_{\vartheta,\Gamma}$ is nonatomic, this number will usually not be less than that. Since we are estimating a distribution, it is clear that the sample size should not be too small, say 100 at least. To compute the exact m.l. estimator numerically is, therefore, troublesome.

Hence one might be willing to settle with as. suboptimal e.s. for ϑ which are easier to compute. Such e.s. may be obtained by distinguishing a certain convex subfamily $\mathcal{G}_0 \subset \mathcal{G}$, and considering the class of functions (see (7.7)) $L_0(\cdot, \vartheta, \Gamma)$, with Γ restricted to $\mathcal{G}_0$.

At first we discuss conditions on the family (8.1') which guarantee that m.l. e.s. are consistent.

Specialized to the case of mixtures the Theorem in Pfanzagl (1989) reads as follows. Applied with $m(\cdot, \vartheta, \alpha)$ replaced by $p_0(S(\cdot, \vartheta), \vartheta, \Gamma)$, the m.l. estimator is defined as a function $\Gamma^{(n)} : X^n \to \mathcal{G}_0$, maximizing

$$\Gamma \to \sum_1^n \log p_0\big(S(x_\nu, \vartheta^{(n)}(\underline{x})), \vartheta^{(n)}(\underline{x}), \Gamma\big) \quad \text{for } \Gamma \in \mathcal{G}_0. \tag{8.4'}$$

Because of the linearity of $\Gamma \to p_0(y, \vartheta, \Gamma)$, this is equivalent to determining $\Gamma^{(n)}(\underline{x}) \in \mathcal{G}_0$ such that

$$n^{-1} \sum_1^n \frac{p_0\big(S(x_\nu, \vartheta^{(n)}(\underline{x})), \vartheta^{(n)}(\underline{x}), \Gamma\big)}{p_0\big(S(x_\nu, \vartheta^{(n)}(\underline{x})), \vartheta^{(n)}(\underline{x}), \Gamma^{(n)}(\underline{x})\big)} \le 1 \quad \text{for } \Gamma \in \mathcal{G}_0. \tag{8.4''}$$

In these expressions, $\vartheta^{(n)}$ is a preliminary e.s. converging under $P^n_{\vartheta_0, \Gamma_0}$ to ϑ_0. A natural choice for $\mathcal{G}_0$ would be the family of all p-measures over $(H, \mathcal{C})$, or an appropriate subfamily. For technical reasons $\mathcal{G}_0$ has to be compact. Hence we may replace $\mathcal{G}_0$ by the family of all sub-p-measures over $(H, \mathcal{C})$, say $\mathcal{G}_*$, which is convex and compact in the topology of vague convergence if H is a locally compact Hausdorff space with countable base. (See Bauer, 1981, p. 243, Corollary 7.8.3.) This is only a technical device. Observe that any sub-p-measure $\Gamma^{(n)}(\underline{x})$ fulfilling (8.4') or the equivalent condition (8.4'') is necessarily a p-measure. Alternatively, we may choose for $\mathcal{G}_0$ the family of all p-measures over $\mathcal{C}$ with a given compact support.

The sequence of m.l. estimators converges under $P^n_{\vartheta_0, \Gamma_0}$ weakly to Γ_0 if $(\vartheta, \Gamma) \to p_0\big(S(x, \vartheta), \vartheta, \Gamma\big)$ is continuous on $\Theta \times \mathcal{G}_*$, for every $x \in X$. The latter holds true if $\eta \to p_0(y, \vartheta, \eta) \in \mathcal{C}_0(H)$ for every $y \in Y$, $\vartheta \in \Theta$, if $(y, \vartheta) \to p_0(y, \vartheta, \eta)$ is equicontinuous in η and if $\vartheta \to S(x, \vartheta)$ is continuous for every $x \in X$. (See Lemma L.21.)

To be applicable in Lemma 7.16, the m.l. estimator ought to be computed with a discretized version in place of $\vartheta^{(n)}(\underline{x})$.

Now we discuss the conditions on the family (8.1') needed for the validity of the improvement procedure (5.7). Aside: Consistency of the preliminary e.s. $\vartheta^{(n)}$, $n \in \mathbb{N}$, was all we needed to obtain a consistent e.s. for Γ. In the improvement procedure, $\sqrt{n}$-consistency is required for $\vartheta^{(n)}$.

To obtain an as. efficient e.s. for ϑ, we apply the improvement procedure (5.7) with $L_0(\cdot, \vartheta, \Gamma)$ given by (7.7). Theorem 5.6, applied for $N = L_0$, requires under (i) that the function $(x, \vartheta, \Gamma) \to L_0^\bullet(x, \vartheta, \Gamma)$ fulfills condition L.5$((\vartheta_0, \Gamma_*), P_{\vartheta_0, \Gamma_0})$. We desist from writing this condition down because it becomes much simpler in the examples to follow than in the case of a general mixture model. Theorem 5.6 requires, moreover, condition (iv). According to Lemma 7.16 and Proposition 7.18, this condition follows for L_0 if for every $\varepsilon > 0$ there exists a weak neighborhood $V_\varepsilon \subset \mathcal{G}_0$ of Γ_* such that

$$\int B(s, \vartheta_n) \sup_{\Gamma \in V_\varepsilon} \left(\frac{p_0'(s, \vartheta_n, \Gamma)}{p_0(s, \vartheta_n, \Gamma)} - \frac{p_0'(s, \vartheta_n, \Gamma_*)}{p_0(s, \vartheta_n, \Gamma_*)} \right)^2 R_{\vartheta_n, \Gamma_0}(ds) < \varepsilon. \qquad (8.5)$$

Finally, Theorem 5.6 requires under (iii) an e.s. $d^{(n)}$ for $P_{\vartheta_0, \Gamma_0}(L_0^\bullet(\cdot, \vartheta_0, \Gamma_*))$. The obvious choice is the sample mean of $L_0^\bullet(\cdot, \vartheta_0, \Gamma_*)$, with ϑ_0 and Γ_* replaced by estimators. Since now the sufficient statistic $S(\cdot, \vartheta)$ is available, this opens the possibility of taking instead the sample mean of $P_{\vartheta, \cdot}^{S(\cdot, \vartheta)}(L_0^\bullet(\cdot, \vartheta, \Gamma))$ (see Remark 7.19).

Recall that the e.s. $\Gamma^{(n)}$ obtained by the m.l. method restricted to $\mathcal{G}_0$ does not necessarily converge to the value of Γ which minimizes the as. variance

$$\Gamma \to P_{\vartheta_0, \Gamma_0}(L_0(\cdot, \vartheta_0, \Gamma)^2) / \left(P_{\vartheta_0, \Gamma_0}(L_0^\bullet(\cdot, \vartheta_0, \Gamma)) \right)^2$$

for $\Gamma \in \mathcal{G}_0$, except for the case $\Gamma_0 \in \mathcal{G}_0$. (See Remark 5.34 and Proposition 9.25.)

If $\mathcal{G}_0 = \mathcal{G}$, we have $\Gamma_* = \Gamma_0$. In this case,

$$P_{\vartheta_0, \Gamma_0}(L_0^\bullet(\cdot, \vartheta_0, \Gamma_*)) = -P_{\vartheta_0, \Gamma_0}(L_0(\cdot, \vartheta_0, \Gamma_0)^2),$$

and this opens another possibility of estimating this quantity, namely by the sample mean of $L_0(\cdot, \vartheta_0, \Gamma_0)^2$ or — preferably — of $P_{\vartheta_0, \cdot}^{S(\cdot, \vartheta_0)}(L_0(\cdot, \vartheta_0, \Gamma_0)^2)$ with ϑ_0, Γ_0 replaced by estimators. Finally, one might in this case also consider replacing ϑ_0, Γ_0 by estimators in $P_{\vartheta_0, \Gamma_0}(L_0^\bullet(\cdot, \vartheta_0, \Gamma_0))$. The consistency of these e.s. requires similar, but slightly different regularity conditions of the original p-measures $P_{\vartheta, \eta}$.

To summarize: Whatever the estimation method for Γ, nothing worse can happen than an increase in the as. variance of the resulting e.s. for ϑ. This increase will be small if $\mathcal{G}_0$ is chosen large enough to contain a p-measure close to Γ for any $\Gamma \in \mathcal{G}$. But circumspection is required in choosing the estimator for $P_{\vartheta_0, \Gamma_0}(L_0^\bullet(\cdot, \vartheta_0, \Gamma_*))$. With a wrong estimator (say $P_{\vartheta^{(n)}, \Gamma^{(n)}}(L_0^\bullet(\cdot, \vartheta^{(n)}, \Gamma^{(n)}))$ if $\Gamma^{(n)}$ is close to $\Gamma_* \neq \Gamma_0$) the estimator for ϑ may be severely biased. See also p. 65.

Let us now consider the following particular case of a parametric subfamily: We choose points $\eta_i \in H$, $i = 1, \ldots, m$, and consider the family of all p–measures with support $\{\eta_1, \ldots, \eta_m\}$, denoting by Γ_α the p–measure

$$\Gamma_\alpha(B) = \sum_1^m \alpha_i 1_B(\eta_i), \qquad B \in \mathcal{C}, \tag{8.6}$$

with $\alpha \in A := \{(\alpha_1, \ldots, \alpha_m) \in [0,1]^m : \sum_1^m \alpha_i = 1\}$. With m not too small (say larger than 20), it might be expensive to estimate the value α_0 which minimizes the as. variance (5.8). Since $\{\Gamma_\alpha : \alpha \in A\}$ is now a convex and weakly compact set of p–measures, the m.l. estimator (based on fixed $\eta_1, \ldots, \eta_m$) converges to the value α_0 which maximizes $\alpha \to R_{\vartheta_0, \Gamma_0}(\log p_0(\cdot, \vartheta_0, \Gamma_\alpha))$ for $\alpha \in A$. If m is sufficiently large and $\eta_1, \ldots, \eta_m$ are appropriately chosen, Γ_{α_0} will be close to Γ_0, hence the improved e.s. for ϑ almost as. efficient. Since we are now working with a finite dimensional subfamily, the m.l. (or any other appropriate) e.s. for α will be $\sqrt{n}$–consistent. Hence Proposition 3.28 applies and there is no need for the discretization–trick.

The Theorem in Pfanzagl (1989) asserts consistency not only for the sequence of m.l. estimators, but for a large class of *asymptotic* m.l. estimators. With this concept one is free to restrict the m.l. estimator for the sample size n to a certain subset $\mathcal{G}_n \subset \mathcal{G}_*$. If H is a separable metric space with a countable dense subset $\{\eta_i : i \in \mathbb{N}\}$, one may choose $\mathcal{G}_n$ as the class of all sub–p–measures with supporting points $\{\eta_1, \ldots, \eta_{m_n}\}$. It follows from Parthasarathy (1967, p. 44, Theorem 6.3) that $\bigcup_1^\infty \mathcal{G}_n$ is dense in $\mathcal{G}_*$ with respect to the vague topology if $m_n \to \infty$. This suffices to guarantee consistency of the as. m.l. e.s.

Nothing can prevent us from interpreting the family of p–measures (8.6) as an element of a sieve, hence the resulting (restricted) m.l. estimator as an element of a sequence of (unrestricted) as. m.l. estimators, which converges to the *true* p–measure. This interpretation would justify the use of $L_0(\cdot, \vartheta, \Gamma)^2$ (or $P_{\vartheta,\cdot}^{S(\cdot,\vartheta)}(L_0(\cdot, \vartheta, \Gamma)^2)$) instead of $L_0^\bullet(\cdot, \vartheta, \Gamma)$ (or $P_{\vartheta,\cdot}^{S(\cdot,\vartheta)}(L_0^\bullet(\cdot, \vartheta, \Gamma))$) in the estimation of $P_{\vartheta,\Gamma}(L_0^\bullet(\cdot, \vartheta, \Gamma))$. Although this is justifiable by as. theorems, one can expect that the use of $L_0^\bullet$ (or $P_{\vartheta,\cdot}^S(L_0^\bullet)$) yields numerically better results.

9. Examples of mixture models

In Section 8 we studied mixture models of the following type: $\{P_{\vartheta,\eta} : \vartheta \in \Theta, \eta \in H\}$ is a family of p–measures with a μ–density fulfilling (8.1), i.e. $p(x,\vartheta,\eta) = q(x,\vartheta)\, p_0(S(x,\vartheta),\vartheta,\eta)$. The observations are from a mixture $P_{\vartheta,\Gamma}$, where Γ is an unknown p–measure on $(H,\mathcal{C})$.

In this section we apply the results of Section 8 to three examples. We start with the common features of these.

The parameter sets Θ and H are $\mathbb{R}_+$ or $\mathbb{R}$, and $S : X \times \Theta \to Y$, with $Y = H$.

(9.1) The sufficient statistic $S(\cdot,\vartheta)$ and the dominating measure μ can be chosen such that
$$p(x,\vartheta,\eta) = A(\eta)\exp[-\eta S(x,\vartheta)].$$

(9.2) $\mu * S(\cdot,\vartheta)$ is dominated by the Lebesgue–measure, for every $\vartheta \in \Theta$.

Properties (9.1) and (9.2) imply in particular (see Remark 7.3) that $R_{\vartheta,\eta}$ ($:= P_{\vartheta,\eta} * S(\cdot,\vartheta)$) has density $p_0(s,\eta) := A(\eta)\exp[-\eta s]$, $s \in Y$, with respect to an appropriately chosen measure ν on $(Y,\mathcal{B})$. As a particular consequence we obtain that $R_{\vartheta,\eta}$ does not depend on ϑ. Hence we write R_η from now on. $\{R_\eta : \eta \in H\}$ is complete (in the sense of (8.3)) provided H has a nonempty interior, which is the case in our examples. This implies $T_0(P_{\vartheta,\Gamma}) = \mathcal{L}_S(P_{\vartheta,\Gamma})$, so that the function $L_0(\cdot,\vartheta,\Gamma)$ defined by (7.7) is the orthogonal component of $\ell^\bullet(\cdot,\vartheta,\Gamma)$ with respect to $T_0(P_{\vartheta,\Gamma})$ and will, therefore, be denoted by $L(\cdot,\vartheta,\Gamma)$ from now on. From (9.1),

$$L(x,\vartheta,\Gamma) = b(x,\vartheta) H_0\big(S(x,\vartheta),\Gamma\big), \tag{9.3}$$

with

$$H_0(s,\Gamma) := -\int \eta A(\eta)\exp[-\eta s]\Gamma(d\eta) \Big/ \int A(\eta)\exp[-\eta s]\Gamma(d\eta), \tag{9.4}$$

and $b(\cdot,\vartheta)$ the orthogonal component of $S^\bullet(\cdot,\vartheta)$ with respect to $\mathcal{L}_S(P_{\vartheta,\Gamma})$ (see (7.6'')).

In the following examples, (9.3) holds with

$$b(x,\vartheta) = S^\bullet(x,\vartheta). \tag{9.5}$$

This relation follows from (9.1), (9.2) under a slight additional regularity condition.

Proposition 9.6. Assume (9.1) and (9.2), and that for every $\eta \in H$, the function $(x,\vartheta) \to S^\bullet(x,\vartheta)e^{-\eta S(x,\vartheta)}$ fulfills $L.5(\vartheta_0, \mu)$. Then

$$P^{S(\cdot,\vartheta_0)}_{\vartheta_0,\cdot}\bigl(S^\bullet(\cdot,\vartheta_0)\bigr) = 0 \qquad R_\eta\text{-a.e.}$$

Proof. For any bounded function h,

$$P_{\vartheta,\eta}\bigl(h(S(\cdot,\vartheta))\bigr) = R_\eta(h).$$

If h has, in addition, a bounded derivative, then

$$\vartheta \to \int h\bigl(S(x,\vartheta)\bigr) A(\eta) e^{-\eta S(x,\vartheta)} \mu(dx) \quad (= P_{\vartheta,\eta}(h(S(\cdot,\vartheta))))$$

may be differentiated under the integral and we obtain

$$\int S^\bullet(x,\vartheta_0)\bigl[h'\bigl(S(x,\vartheta_0)\bigr) - \eta h\bigl(S(x,\vartheta_0)\bigr)\bigr] A(\eta) e^{-\eta S(x,\vartheta_0)} \mu(dx) = 0.$$

With $E(s,\vartheta_0) \in P^{S(\cdot,\vartheta_0)=s}_{\vartheta_0,\cdot}\bigl(S^\bullet(\cdot,\vartheta_0)\bigr)$ this implies

$$\int E(s,\vartheta_0)[h'(s) - \eta h(s)] R_\eta(ds) = 0.$$

According to Remark 7.3 R_η has a λ–density $s \to q_0(s) A(\eta) e^{-\eta s}$, so that

$$\int E(s,\vartheta_0) q_0(s) [h'(s) - \eta h(s)] e^{-\eta s} ds = 0.$$

If for some function $f : Y \to \mathbb{R}$

$$\int f(s)[h'(s) - \eta h(s)] e^{-\eta s} ds = 0 \qquad (9.7)$$

holds for all $\eta \in H$ and all bounded functions h with bounded derivative, this implies $f = 0$ λ-a.e.

To see this, apply (9.7) with $h_a(s) := \frac{1}{\sqrt{2\pi}} \int_0^s e^{-\frac{1}{2}(t-a)^2} dt$ $(a \in \mathbb{R})$ and $\eta = 0$ in the case $H = \mathbb{R}$ and $h_\eta(s) = e^{-s\eta}$ in the case $H = \mathbb{R}_+$ so that we obtain $N_{(a,1)}(f) = 0$ $(a \in \mathbb{R})$ and $\int_0^\infty f(s) e^{-2s\eta} ds = 0$ $(\eta \in \mathbb{R}_+)$, respectively. This implies $f = 0$ λ-a.e. by the Completeness Theorem. $\square$

The as. variance bound is (see (7.9))

$$\sigma_0^2(\vartheta, \Gamma) = 1/P_{\vartheta,\Gamma}\bigl(S^\bullet(\cdot,\vartheta)^2 H_0(S(\cdot,\vartheta),\Gamma)^2\bigr). \qquad (9.8)$$

In the examples to follow, it turns out that $P_{\vartheta,\cdot}^{S(\cdot,\vartheta)}(S^\bullet(\cdot,\vartheta)^2)$ is of the form $C(\vartheta)B(s)$ (see (9.10″) and (9.12′)). In these cases, the as. variance bound (9.8) may be written as

$$\sigma_0^2(\vartheta,\Gamma) = \frac{1}{C(\vartheta)} \cdot \hat{\sigma}_0^2(\Gamma), \text{ with } \hat{\sigma}_0^2(\Gamma) = \left(R_\Gamma\left(BH_0(\cdot,\Gamma)^2\right)\right)^{-1}. \tag{9.8′}$$

Remark 9.9. In the examples to follow, $\{R_\eta : \eta \in H\}$ is either a scale–parameter family of Gamma distributions (i.e. $R_\eta = \Gamma_{\eta,b}$), or the location parameter family of normal distributions (i.e. $R_\eta = N_{(-\eta,1)}$).

It is more than pure chance that just these two families occur. The underlying reason is that in all examples the families of p–measures are generated by a transformation group. More precisely, for any fixed $\vartheta \in \Theta$, there exists a group of transformations $T_{\vartheta,\lambda} : X \to X$, $\lambda \in H$, such that, for any $\eta_0 \in H$, $P_{\vartheta,\eta_0} * T_{\vartheta,\lambda}$, $\lambda \in H$, generates the whole family $\{P_{\vartheta,\eta} : \eta \in H\}$. If such a family is 1–dimensional exponential (as required by (9.1)), then the sufficient statistic $S(\cdot,\vartheta)$ can always be chosen such that

$$S(T_{\vartheta,\lambda}x,\vartheta) = \lambda S(x,\vartheta) \quad \text{or} \quad S(T_{\vartheta,\lambda}x,\vartheta) = S(x,\vartheta) + \lambda,$$

and that $R_\eta = P_{\vartheta,\eta} * S(\cdot,\vartheta)$ is Gamma or normal, respectively. (This is a theorem of Dynkin. See Borges and Pfanzagl (1965) for a proof under weak regularity conditions.)

In the following computations we need expressions for the conditional expectations of $S^{\bullet\bullet}(\cdot,\vartheta)$ and $S^\bullet(\cdot,\vartheta)^2$.

Let

$$\hat{C}(s,\vartheta) \in P_{\vartheta,\cdot}^{S(\cdot,\vartheta)=s}(S^{\bullet\bullet}(\cdot,\vartheta)), \tag{9.10′}$$

$$C(s,\vartheta) \in P_{\vartheta,\cdot}^{S(\cdot,\vartheta)=s}(S^\bullet(\cdot,\vartheta)^2). \tag{9.10″}$$

The same type of reasoning as in Proposition 9.6 (starting from $P_{\vartheta,\Gamma}\left(S^\bullet(\cdot,\vartheta)h(S(\cdot,\vartheta))\right) \equiv 0$) shows that

$$\hat{C}(s,\vartheta) = C'(s,\vartheta) + C(s,\vartheta)\frac{q_0'(s)}{q_0(s)}. \tag{9.11}$$

In the examples to follow we have

$$C(s,\vartheta) = C(\vartheta)B(s), \tag{9.12′}$$

with $\vartheta \to C(\vartheta)$ continuous. This implies that

$$\hat{C}(s,\vartheta) = C(\vartheta)\hat{B}(s), \tag{9.12''}$$

with

$$\hat{B}(s) = B'(s) + B(s)\frac{q_0'(s)}{q_0(s)}. \tag{9.13}$$

* * *

Now we turn to conditions which guarantee that Theorem 5.6 holds, i.e. that the improvement procedure (5.7) is applicable. The following result refers to improvement procedures based on functions

$$N(x,\vartheta,\alpha) = S^{\bullet}(x,\vartheta)H\bigl(S(x,\vartheta),\alpha\bigr), \quad \alpha \in A, \tag{9.14}$$

where A is a Hausdorff space with countable base, and $H(\cdot,\alpha)$ an arbitrary function for which $S^{\bullet}(\cdot,\vartheta)^2 H\bigl(S(\cdot,\vartheta),\alpha\bigr)^2$ is $P_{\vartheta,\Gamma}$–integrable for $\alpha \in A$. Because of (9.5), this is a special case of the model specified in (7.13). For $A = \mathcal{G}$ (the family of all p–measures over $(H,\mathcal{C})$) and $H(\cdot,\alpha) = H_0(\cdot,\Gamma)$ (see (9.4)) the resulting improvement procedure leads to as. efficient e.s. (provided conditions (9.1) and (9.2) hold true).

In the following Proposition, α_* is an arbitrary element of A.

Proposition 9.15. *Assume that*

condition $L.5(\vartheta_0, P_{\vartheta_0,\Gamma_0})$ *is fulfilled for*

(i) $(x,\vartheta) \to S^{\bullet\bullet}(x,\vartheta)H\bigl(S(x,\vartheta),\alpha_*\bigr)$,

(ii) $(x,\vartheta) \to S^{\bullet}(x,\vartheta)^2 H'\bigl(S(x,\vartheta),\alpha_*\bigr)$

and that

condition $L.5(\alpha_*, R_{\Gamma_0})$ *is fulfilled for*

(iii) $(s,\alpha) \to \hat{B}(s)H(s,\alpha)$,

(iv) $(s,\alpha) \to B(s)H'(s,\alpha)$,

(v) $(s,\alpha) \to B(s)H(s,\alpha)^2$.

Assume we are given e.s. $\vartheta^{(n)} : X^n \to \Theta$ *and* $\hat{\alpha}^{(n)} : Y^n \to A$ *with the following properties:*

(vi) $\vartheta^{(n)} = \vartheta_0 + O_p(n^{-1/2}) \quad (P^n_{\vartheta_0,\Gamma_0})$,

(vii) $\hat{\alpha}^{(n)} = \alpha_* + o_p(n^0)$ $(R^n_{\Gamma_0})$.

Define $\hat{\vartheta}^{(n)}$ by

$$\hat{\vartheta}^{(n)}(\underline{x}) := \vartheta^{(n)}_*(\underline{x}) \qquad (9.16)$$

$$- \frac{\sum_1^n S^\bullet(x_\nu, \vartheta^{(n)}_*(\underline{x})) H\big(S(x_\nu, \vartheta^{(n)}_*(\underline{x})), \hat{\alpha}^{(n)}\big(S(x_1, \vartheta^{(n)}_*(\underline{x})), \ldots, S(x_n, \vartheta^{(n)}_*(\underline{x})) \big)\big)}{C(\vartheta^{(n)}_*(\underline{x})) \sum_1^n D\big(S(x_\nu, \vartheta^{(n)}_*(\underline{x})), \hat{\alpha}^{(n)}\big(S(x_1, \vartheta^{(n)}_*(\underline{x})), \ldots, S(x_n, \vartheta^{(n)}_*(\underline{x})) \big)\big)}$$

where $\vartheta^{(n)}_*$ is a discretized version of $\vartheta^{(n)}$, and

$$D(s, \alpha) := \hat{B}(s) H(s, \alpha) + B(s) H'(s, \alpha). \qquad (9.17)$$

Then $\hat{\vartheta}^{(n)}$, $n \in \mathbb{N}$, is under $P^n_{\vartheta_0, \Gamma_0}$ as. normal with mean 0 and variance

$$\sigma^2(\vartheta_0, \Gamma_0, \alpha_*) = \frac{1}{C(\vartheta_0)} \hat{\sigma}^2(\Gamma_0, \alpha_*), \qquad (9.18)$$

with

$$\hat{\sigma}^2(\Gamma_0, \alpha_*) := \frac{R_{\Gamma_0}(BH(\cdot, \alpha_*)^2)}{\big(R_{\Gamma_0}(D(\cdot, \alpha_*))\big)^2}. \qquad (9.18')$$

Addendum. If $A = \mathcal{G}$, $H(\cdot, \alpha) = H_0(\cdot, \Gamma)$ (see (9.4)) and $\alpha_* = \Gamma_0$, then $\hat{\sigma}^2(\Gamma_0, \alpha_*)$ becomes

$$\hat{\sigma}^2(\Gamma_0, \Gamma_0) = \big(R_{\Gamma_0}(BH_0(\cdot, \Gamma_0)^2)\big)^{-1},$$

which agrees with the factor of the as. variance bound, $\hat{\sigma}_0^2(\Gamma_0)$, given in (9.8').

Since

$$P_{\vartheta, \Gamma}\big(L(\cdot, \vartheta, \Gamma)^2\big) = -P_{\vartheta, \Gamma}\big(L^\bullet(\cdot, \vartheta, \Gamma)\big), \qquad (9.19)$$

this follows immediately from the fact that $C(\vartheta) D(\cdot, \Gamma_0)$ is a version of

$$P^{S(\cdot, \vartheta)}_{\vartheta, \cdot}\big(L^\bullet(\cdot, \vartheta, \Gamma_0)\big).$$

According to Proposition 3.28 the use of discretized estimators in (9.16) becomes superfluous if $\hat{\alpha}^{(n)} = \alpha_* + O_p(n^{-1/2})$ (which will usually be the case if $A \subset \mathbb{R}^k$).

Proof. The assertion follows from Proposition 5.6, applied for $N(\cdot, \vartheta, \alpha)$ defined by (9.14).

(i) Condition 5.6(i) follows from 9.15(i),(ii).

(ii) With $\mathcal{C}(\vartheta)$ continuous, condition 5.6(iii) for

$$d^{(n)}(\underline{x}) = C(\vartheta_*^{(n)}(\underline{x})) n^{-1} \sum_1^n D\Big(S(x_\nu, \vartheta_*^{(n)}(\underline{x})),$$

$$\hat{\alpha}^{(n)}\big(S(x_1, \vartheta_*^{(n)}(\underline{x})), \ldots, S(x_n, \vartheta_*^{(n)}(\underline{x}))\big)\Big)$$

follows from Proposition L.9, since conditions 9.15(iii) and (iv) imply L.5(α_*, R_{Γ_0}) for the function $(s, \alpha) \to D(s, \alpha)$. Observe that $C(\vartheta) D(\cdot, \alpha)$ is an element of $P_{\vartheta, \cdot}^{S(\cdot, \vartheta)}(N^\bullet(\cdot, \vartheta, \alpha))$.

(iii) Condition 9.15(v) implies condition L.5(α_*, Γ_0) for

$$(s, \alpha) \to B(s)\big(H(s, \alpha) - H(s, \alpha_*)\big)^2.$$

By Lemma L.7 this implies for every $\varepsilon > 0$ the existence of a neighborhood $V_\varepsilon \ni \alpha_*$ such that

$$\int B(s) \sup_{\alpha \in V_\varepsilon} \big(H(s, \alpha) - H(s, \alpha_*)\big)^2 R_{\Gamma_0}(ds) < \varepsilon.$$

By Proposition 7.18 and Lemma 7.16 this implies condition 5.6(iv) with $N_\nu^{(n)}(\vartheta, (x_1, \ldots, x_n))$ and $N(x_\nu, \vartheta, \alpha_0)$ replaced by

$$S^\bullet(x_\nu, \vartheta) H\big(S(x_\nu, \vartheta), \hat{\alpha}^{(n)}(S(x_1, \vartheta), \ldots, S(x_n, \vartheta))\big)$$

and

$$S^\bullet(x_\nu, \vartheta) H\big(S(x_\nu, \vartheta), \alpha_*\big),$$

respectively. □

Remark 9.20.
Condition L.5$((\vartheta_0, \alpha_*), P_{\vartheta_0, \Gamma_0})$ for a function $(x, \vartheta, \alpha) \to e(x, \vartheta) F(S(x, \vartheta), \alpha)$ implies condition L.5(α_*, R_{Γ_0}) for the function $(s, \alpha) \to E(s, \vartheta_0) F(s, \alpha)$, with $E(\cdot, \vartheta_0) \in P_{\vartheta_0, \cdot}^{S(\cdot, \vartheta_0)}(e(\cdot, \vartheta_0))$.

To see this, we remark that for any function $f : X \times Y \to \mathbb{R}$, condition L.(5' & 5") is equivalent to condition (L.5') for $|f|$. For any $V \ni \vartheta_0$

$$R_{\Gamma_0}\big(\sup_{\alpha \in U} |E(\cdot, \vartheta_0) F(\cdot, \alpha)|\big)$$

$$= R_{\Gamma_0}\big(|E(\cdot, \vartheta_0)| \sup_{\alpha \in U} |F(\cdot, \alpha)|\big)$$

$$\leq P_{\vartheta_0, \Gamma_0}\big(|e(\cdot, \vartheta_0)| \sup_{\alpha \in U} |F(S(\cdot, \vartheta_0), \alpha)|\big)$$

$$\leq P_{\vartheta_0, \Gamma_0}\big(\sup_{\vartheta \in V} \sup_{\alpha \in U} |e(\cdot, \vartheta) F(S(\cdot, \vartheta), \alpha)|\big).$$

Therefore, condition L.5$((\vartheta_0, \alpha_*), P_{\vartheta_0,\Gamma_0})$ for

$$(x, \vartheta, \alpha) \to S^{\bullet\bullet}(x, \vartheta) H(S(x, \vartheta), \alpha), \qquad (9.21)$$

$$(x, \vartheta, \alpha) \to S^{\bullet}(x, \vartheta)^2 H'(S(x, \vartheta), \alpha) \qquad (9.22)$$

implies conditions 9.15(i),(iii) and 9.15(ii),(iv), respectively.

Conditions 9.15(i),(ii),(iii),(iv),(v) (as well as conditions (9.21), (9.22)) refer to the value α_* to which $\alpha^{(n)}$, $n \in \mathbb{N}$, converges under $P^n_{\vartheta_0,\Gamma_0}$. To establish the validity of the improvement procedure (9.16) it therefore suffices to check these conditions for those values $\alpha_* \in A$ which possibly occur as limits. This distinction is relevant because the application of consistency theorems usually requires to work with compact sets A as a technical device, whereas the limits of e.s. $\alpha^{(n)}$, $n \in \mathbb{N}$, are confined to some subset of A. As a particular example, we mention the model where Γ may be any p-measure on $(H, \mathcal{C})$. A natural choice for A would be $\mathcal{G}$, the class of all p-measures on $(H, \mathcal{C})$. Since $\mathcal{G}$ fails to be compact with respect to the vague topology, we resort to $\mathcal{G}_*$, the class of all sub-p-measures on $(H, \mathcal{C})$, for establishing the consistency of sequences of m.l. estimators. The limits of such sequences are, however, p-measures, hence in $\mathcal{G}$. Whenever this holds true, it suffices to check the conditions of Proposition 9.15 for $\Gamma_* \in \mathcal{G}$.

In applications with $A = \mathcal{G}$ and $H(\cdot, \alpha) = H_0(\cdot, \Gamma)$ there is another small simplification. Conditions (9.22) and 9.15(v) follow if condition L.5$((\vartheta_0, \Gamma_*), P_{\vartheta_0,\Gamma_0})$ holds for

$$(x, \vartheta, \Gamma) \to S^{\bullet}(x, \vartheta)^2 H_2(S(x, \vartheta), \Gamma), \qquad (9.23)$$

with

$$H_2(s, \Gamma) := \int \eta^2 A(\eta) \exp[-\eta s] \Gamma(d\eta) / \int A(\eta) \exp[-\eta s] \Gamma(d\eta). \qquad (9.24)$$

This follows immediately from the fact that $H_0(s,\Gamma)^2 \leq H_2(s,\Gamma)$ and $H'_0(s,\Gamma) = H_2(s,\Gamma) - H_0(s,\Gamma)^2$, hence $0 \leq H'_0(s,\Gamma) \leq H_2(s,\Gamma)$.

Hence it suffices in this case to check conditions (9.21) and (9.23) for $\alpha_* = \Gamma_* \in \mathcal{G}$.

$$\bullet \qquad * \qquad *$$
$$*$$

Proposition 9.15 presumes under (vii) the existence of an e.s. $\hat{\alpha}^{(n)}$, $n \in \mathbb{N}$, converging to α_* stochastically. Hence the application for $A = \mathcal{G}$, $H(\cdot, \alpha) =$

$H_0(\cdot, \Gamma)$ requires an e.s. $\hat{\Gamma}^{(n)}|Y^n$, $n \in \mathbb{N}$, converging under $R_{\Gamma_0}^n$ to some Γ_* (preferably one close to Γ_0). The following Proposition 9.25 ascertains that such e.s. can be obtained by the m.l. method.

Let $\mathcal{G}_0$ be a convex and compact subset of $\mathcal{G}_*$ (the family of all sub–p–measures on $(H, \mathcal{C})$) with the property that $\Gamma \in \mathcal{G}_0$ implies $\Gamma/\Gamma(Y) \in \mathcal{G}_0$ for $\Gamma(Y) \neq 0$.

Proposition 9.25. *Assume conditions (9.1) and (9.2).*

a) For any p–measure $\Gamma_0 \in \mathcal{G}$ there exists a p–measure $\Gamma_* \in \mathcal{G}_0$ such that

$$R_{\Gamma_0}\left(\log \frac{p_0(\cdot, \Gamma_*)}{p_0(\cdot, \Gamma)}\right) > 0 \quad \text{for } \Gamma \in \mathcal{G}_0, \ \Gamma \neq \Gamma_*. \tag{9.26}$$

Observe that $\Gamma_0 \in \mathcal{G}_0$ implies $\Gamma_* = \Gamma_0$.

b) Let $\mathcal{G}_n \subset \mathcal{G}_0$, $n \in \mathbb{N}$, be an increasing sequence of convex and compact subsets with the property that $\Gamma \in \mathcal{G}_n$ implies $\Gamma/\Gamma(Y) \in \mathcal{G}_n$. Assume that $\bigcup_1^\infty \mathcal{G}_n$ is dense in $\mathcal{G}_0$. Let $\hat{\Gamma}^{(n)} \in \mathcal{G}_n$, $n \in \mathbb{N}$, be an e.s. which is as. m.l. in the following sense.

There exists a finite $c \geq 0$ such that for all $n \in \mathbb{N}$, $(s_1, \ldots, s_n) \in Y^n$, $\Gamma \in \mathcal{G}_n$

$$\sum_1^n \log \frac{p_0(s_\nu, \hat{\Gamma}^{(n)}(s_1, \ldots, s_n))}{p_0(s_\nu, \Gamma)} \geq -c. \tag{9.27}$$

Then
$$\hat{\Gamma}^{(n)} \Rightarrow \Gamma_* \quad (R_{\Gamma_0}^n).$$

Proof. a) Follows from the Proposition in Pfanzagl (1989), applied with $A = \mathcal{G}_0$, $P_0 = R_{\Gamma_0}$ and $m(s, \Gamma) = p_0(s, \Gamma)/p_0(s, \Gamma_0)$. Since $p_0(s, \eta) \in \mathcal{C}_0(H)$ for ν–a.a. $s \in Y$, the function $\Gamma \to p_0(s, \Gamma)$ is continuous with respect to the vague topology. Moreover,

$$R_{\Gamma_0}\left(\log[p_0(\cdot, \Gamma)/p_0(\cdot, \Gamma_0)]\right) \leq 0.$$

Hence the Proposition applies. The inequality in (9.26) is strict since Γ_* is identifiable by Proposition 6.2 in Pfanzagl (1988).

b) Follows immediately from the Theorem in Pfanzagl (1989). □

* * *

Finally, we also investigated the performance of the improved estimators for ϑ if the estimator for Γ is restricted to a rather small parametric family $\mathcal{G}_0$. This was done in Examples 1 and 2, where $H = (0, \infty)$, taking for $\mathcal{G}_0$ the family of all Gamma distributions (including a scale parameter). Since this family contains a great variety of (unimodal, though) distributions, one might hope that the loss of efficiency will be small, at least if the true mixing distribution is unimodal.

Let $\Gamma_{\alpha,\beta}$ denote the Gamma distribution with λ–density

$$\eta \to \frac{\alpha^\beta}{\Gamma(\beta)} \eta^{\beta-1} \exp[-\alpha\eta], \qquad \eta > 0. \tag{9.28}$$

For a model (9.1) with $A(\eta) = C_k \eta^k$ we obtain that the λ–density of $R_{\Gamma_{\alpha,\beta}}$ is

$$s \to \frac{\Gamma(\beta+k)}{\Gamma(\beta)\Gamma(k)} \cdot \frac{\alpha^\beta s^{k-1}}{(\alpha+s)^{\beta+k}}, \qquad s > 0.$$

Moreover,

$$H_0(s, \Gamma_{\alpha,\beta}) = -\frac{\beta+k}{\alpha+s}, \qquad s > 0.$$

Since the shape parameter β enters this expression only through a factor not depending on s, we may choose for H in (9.14)

$$H(s, \alpha) = -(\alpha+s)^{-1}, \qquad s > 0. \tag{9.29}$$

Since this function $H(\cdot, \alpha)$ is proportional to $H_0(\cdot, \Gamma_{\alpha,\beta})$ for any $\beta > 0$, we can expect to obtain high as. efficiency for a large class of distributions Γ (and as. efficiency 1 for *all* Gamma distributions).

In principle, we could take for α any value in $(0, \infty)$, thus obtaining from (9.16) an e.s. for ϑ_0 with as. variance $\sigma^2(\vartheta_0, \Gamma_0, \alpha)$ given by (9.18). To make this as. variance as small as possible, it was suggested in Section 5 to determine the e.s. for α by minimizing the sample analogue of (5.8) (see (5.29)). The presence of a sufficient statistic $S(\cdot, \vartheta)$ opens an alternative to the minimization of (5.29), namely the minimization of the sample analogue of (9.18′).

For $H(\cdot, \alpha)$ given by (9.29) we have (see (9.17))

$$D(s, \alpha) = \frac{B(s)}{(\alpha+s)^2} - \frac{\hat{B}(s)}{\alpha+s}. \tag{9.30}$$

Hence we use the e.s.

$$\alpha^{(n)}(x_1, \ldots, x_n) := \hat{\alpha}^{(n)}\left(S(x_1, \vartheta^{(n)}(\underline{x})), \ldots, S(x_n, \vartheta^{(n)}(\underline{x}))\right),$$

where $\hat{\alpha}(s_1,\ldots,s_n)$ is determined as the solution in α of

$$\sum_1^n \frac{B(s_\nu)}{(\alpha+s_\nu)^2}\Big(\sum_1^n\big(\frac{\hat{B}(s_\nu)}{\alpha+s_\nu} - \frac{B(s_\nu)}{(\alpha+s_\nu)^2}\big)\Big)^{-2} = \min. \qquad (9.31)$$

This leads to an improved e.s. for ϑ_0 with as. variance (see (9.18))

$$\sigma^2(\vartheta_0,\Gamma_0,\alpha_0) = \frac{1}{C(\vartheta_0)}\hat{\sigma}^2(\Gamma_0,\alpha_0), \quad \text{with} \qquad (9.32)$$

$$\hat{\sigma}^2(\Gamma_0,\alpha_0) := \int \frac{B(s)}{(\alpha_0+s)^2} R_{\Gamma_0}(ds)\Big(\int \big(\frac{\hat{B}(s)}{\alpha_0+s} - \frac{B(s)}{(\alpha_0+s)^2}\big) R_{\Gamma_0}(ds)\Big)^{-2},$$

where α_0 minimizes $\alpha \to \hat{\sigma}^2(\Gamma_0,\alpha)$ for $\alpha \in (0,\infty)$.

To infer convergence of $\alpha^{(n)}$, $n \in \mathbb{N}$, to α_0, we apply the Consistency–Lemma L.33 with $A = [0,\infty]$ and $N_i(x,\vartheta,\alpha) = \tilde{N}_i\big(S(x,\vartheta),\alpha\big)$, where

$$\tilde{N}_1(s,\alpha) = (1+\alpha)\hat{B}(s)/(\alpha+s), \qquad (9.33)$$
$$\tilde{N}_2(s,\alpha) = (1+\alpha)^2 B(s)/(\alpha+s)^2. \qquad (9.34)$$

The factors $(1+\alpha)$ and $(1+\alpha)^2$, respectively, enable us to consider $N_i(\cdot,\vartheta,\alpha)$ for α in the *compact* set $[0,\infty]$, which is required in the Consistency–Lemma. Condition L.5$((\vartheta_0,\beta),P_{\vartheta_0,\Gamma_0})$ for $\beta \in [0,\infty]$ follows for $N_i(\cdot,\vartheta,\alpha)$, $i = 1,2$, from condition L.5$(\vartheta_0,P_{\vartheta_0,\Gamma_0})$ for

$$(x,\vartheta) \to \hat{B}\big(S(x,\vartheta)\big)/S(x,\vartheta), \quad (x,\vartheta) \to \hat{B}\big(S(x,\vartheta)\big) \qquad (9.35)$$

and

$$(x,\vartheta) \to B\big(S(x,\vartheta)\big)/S(x,\vartheta)^2, \quad (x,\vartheta) \to B\big(S(x,\vartheta)\big). \qquad (9.36)$$

The Consistency–Lemma presumes, moreover, the existence of a *unique* α_0 minimizing $\alpha \to \hat{\sigma}^2(\Gamma_0,\alpha)$ for $\alpha \in (0,\infty)$. Because of the compactness condition for A, this has to be supplemented by the condition that $\hat{\sigma}^2(\Gamma_0,\alpha)$ remains bounded away from $\hat{\sigma}^2(\Gamma_0,\alpha_0)$ as α tends to 0 or ∞. This condition is certainly fulfilled if Γ_0 happens to be a Gamma distribution. (We have $\hat{\sigma}^2(\Gamma_{\alpha_0,\beta_0},\alpha) \geq \hat{\sigma}_0^2(\Gamma_{\alpha_0,\beta_0})$. Rewriting $\hat{\sigma}^2(\Gamma_{\alpha_0,\beta_0},\alpha)$ as

$$P_{\vartheta_0,\Gamma_{\alpha_0,\beta_0}}\big((1+\alpha)^2 N(\cdot,\vartheta_0,\alpha)^2\big)/\big(P_{\vartheta_0,\Gamma_{\alpha_0,\beta_0}}((1+\alpha)N(\cdot,\vartheta_0,\alpha)L(\cdot,\vartheta_0,\Gamma_{\alpha_0,\beta_0}))\big)^2,$$

it follows from Remark 5.30 that $\hat{\sigma}^2(\Gamma_{\alpha_0,\beta_0},\alpha) = \hat{\sigma}_0^2(\Gamma_{\alpha_0,\beta_0})$ for some $\alpha \in [0,\infty]$ iff $(1+\alpha)N(\cdot,\vartheta_0,\alpha) = (1+\alpha)S^\bullet(x,\vartheta_0)/(\alpha+S(x,\vartheta_0))$ is proportional to $L(\cdot,\vartheta_0,\Gamma_{\alpha_0,\beta_0})$. This holds true iff $\alpha = \alpha_0$ (if there exist $x_1,x_2 \in X$ with $S^\bullet(x_i,\vartheta_0) \neq 0$, $i = 1,2$, and $S(x_1,\vartheta_0) \neq S(x_2,\vartheta_0)$, which is the case in our examples).

For later use we put on record that (see (9.18') and (9.8'))

$$\inf_{\alpha>0} \hat{\sigma}^2(\Gamma_{\alpha_0,\beta_0},\alpha) = \hat{\sigma}^2(\Gamma_{\alpha_0,\beta_0},\alpha_0) = \hat{\sigma}_0^2(\Gamma_{\alpha_0,\beta_0}). \qquad (9.37)$$

We leave open the question whether the minimizing value α_0 is unique for arbitrary Γ_0.

To make sure that Proposition 9.15 applies, it suffices to check conditions (9.21), (9.22) and 9.15(v). For the function $H(\cdot, \alpha)$ defined by (9.29) we have $H'(\cdot, \alpha) = H(\cdot, \alpha)^2$. Hence it suffices to establish condition L.5($\vartheta_0, P_{\vartheta_0, \Gamma_0}$) for the functions (9.35), (9.36) and the functions

$$(x, \vartheta) \to S^{\bullet\bullet}(x, \vartheta)/S(x, \vartheta), \qquad (9.38)$$
$$(x, \vartheta) \to S^{\bullet}(x, \vartheta)^2/S(x, \vartheta)^2. \qquad (9.39)$$

For $\beta \in (0, \infty)$, these conditions imply condition L.5$((\vartheta_0, \beta), P_{\vartheta_0, \Gamma_0})$ for the functions $(x, \vartheta, \alpha) \to S^{\bullet\bullet}(x, \vartheta)/(\alpha + S(x, \vartheta))$ and $(x, \vartheta, \alpha) \to S^{\bullet}(x, \vartheta)^2/(\alpha + S(x, \vartheta))^2$, i.e. conditions (9.21) and (9.22).

*   *

*

The reliability of as. theory in this area was examined by simulation experiments. For each of the estimators under consideration (preliminary as well as improved ones), the actual performance in simulation experiments was compared with the results of as. theory. The performance of each estimator, say $\vartheta^{(n)}$, was measured by its bias, $\vartheta^{(n)} - \vartheta_0$, and its standardized mean deviation, $n^{1/2}|\vartheta^{(n)} - \vartheta_0|$. The average of these quantities over the N simulation experiments is to be set against the theoretical values.

For presenting the results we use the following tables.

estimator	bias	mean deviation			coverage
		theor.	empir.	e.m.	sample

In case of the bias, the empirical value is $N^{-1} \sum_{i=1}^{N} \vartheta^{(n)}(\underline{x}_i) - \vartheta_0$ (where $\underline{x}_i$ is the sample obtained in the i-th simulation experiment). The theoretical value is zero. To evaluate the relevance of the bias, it has to be seen in relation to the random error of the estimator. Therefore, we give in the column "bias" not the bias of $\vartheta^{(n)}$ as estimated by the simulation experiment, but its relation to the 99%–error bound of $\vartheta^{(n)}$, computed as $2.58\, s_*$, where s_*^2 is the variance between $\vartheta^{(n)}(\underline{x}_i)$, $i = 1, \ldots, N$. It turns out that in all examples the bias is negligible compared with the random error.

In case of the standardized mean deviation, the empirical value is

$$N^{-1} \sum_{i=1}^{N} n^{1/2}|\vartheta^{(n)}(\underline{x}_i) - \vartheta_0|.$$

The theoretical value is computed as $\sqrt{2/\pi}\sigma(\vartheta_0, \Gamma_0, \alpha_*)$, where $\sigma^2(\vartheta_0, \Gamma_0, \alpha_*)$ is the as. variance of the improved e.s. as given by (9.18).

The main part of the results refer to the model with $\mathrm{A} = \mathcal{G}$, $H(\cdot, \alpha) = H_0(\cdot, \Gamma)$. In this case, we have to compute $\sigma^2(\vartheta_0, \Gamma_0, \Gamma_*)$, where Γ_* is the p–measure to which the e.s. $\Gamma^{(n)}$, $n \in \mathbb{N}$, converges. If $\Gamma^{(n)}$, $n \in \mathbb{N}$, is consistent, then $\Gamma_* = \Gamma_0$ and no problem arises. In general, however, Γ_* depends on the estimation procedure for Γ. The m.l. estimator for instance, if restricted to some subfamily $\mathcal{G}_0 \subset \mathcal{G}_*$, converges to that value Γ_* which maximizes $\Gamma \to R_{\Gamma_0} \log(p_0(\cdot, \Gamma))$ for $\Gamma \in \mathcal{G}_0$ (see Proposition 9.25.a) (and which agrees with Γ_0 only if $\Gamma_0 \in \mathcal{G}_0$).

To assess the deviation of the empirical mean deviation from its theoretical value, we give, in the column "e.m.", the 99%–error margin for the empirical values, computed from the variance between the N simulation results (i.e. $2.58\, s/\sqrt{N}$, where s^2 is the variance between $n^{1/2}|\vartheta^{(n)}(\underline{x}_i) - \vartheta_0|$, $i = 1, \ldots, N$).

The statistician who applies the improvement procedure in practice neither knows the theoretical mean deviation of $\vartheta^{(n)}$, nor its empirical value (obtained from a large number of simulation experiments). After having observed one sample, he will compute the standard deviation $\sigma(\vartheta_0, \Gamma_0, \Gamma_*)$, with Γ_* replaced by the estimate obtained from this sample, and with R_{Γ_0}–integrations replaced by the corresponding sample means.

The average of these standard deviations over the N samples is converted into an estimate for the mean deviation by multiplication with $\sqrt{2/\pi}$, to make it comparable with the figures under "theor." and "empir.". It is listed in the tables under "sample".

To make these results transparent, the numbers under the headline "mean deviation" are not given in their absolute value, but in their relation to the mean deviation corresponding to the minimal as. variance.

Finally, the statistician will use his estimate of the standard deviation to compute a confidence interval for ϑ. We present under "coverage" the relative frequencies with which the symmetric 0.9–confidence interval covers the true value ϑ_0.

In the numerical examples to follow, a rather crude approximation to the m.l. estimator is used. About 10–15 supporting points are chosen in the "range" of Γ (say between the .1 and .9–quantile). Starting from $S(x_\nu, \vartheta^{(n)}(\underline{x}))$, $\nu = 1, \ldots, n$, the E.M. algorithm is applied to compute probabilities for these supporting points which maximize the likelihood.

The results for these estimators are listed under **improved A**. In spite of the fact that the number of supporting points is rather small, and the m.l. estimator approximates the value Γ_* maximizing $\Gamma \to R_{\Gamma_0}(\log p_0(\cdot, \Gamma))$, and does, therefore, not necessarily minimize $\Gamma \to \sigma^2(\vartheta_0, \Gamma_0, \Gamma)$ over the family of p–measures with the given supporting points, the difference between $\sigma(\vartheta_0, \Gamma_0, \Gamma_*)$

and $\sigma(\vartheta_0, \Gamma_0, \Gamma_0)$ is negligible in all examples.

If we consider $\Gamma^{(n)}$ as an approximation to Γ_0 (rather than Γ_*), this would even justify an improvement procedure different from (9.16). To describe this alternative improvement procedure in a convenient way, we introduce

$$\hat{\vartheta}^{(n)}(\underline{x}, \vartheta, \Gamma) := \vartheta - \frac{n^{-1} \sum_1^n S^\bullet(x_\nu, \vartheta) H(S(x_\nu, \vartheta), \Gamma)}{C(\vartheta) R_\Gamma(D(\cdot, \Gamma))} \, . \qquad (9.40)$$

The alternative improvement procedure consists in replacing (ϑ, Γ) in $\hat{\vartheta}^{(n)}(\underline{x}, \vartheta, \Gamma)$ by estimates. This means, in particular, that $R_\Gamma(D(\cdot, \Gamma))$ is replaced by $R_{\Gamma^{(n)}(\underline{x})}(D(\cdot, \Gamma^{(n)}(\underline{x})))$.

The use of this estimate is justified only if $\Gamma^{(n)}(\underline{x})$ is close to Γ_0. If $\Gamma^{(n)}$, $n \in \mathbb{N}$, converges to some $\Gamma_* \neq \Gamma_0$, we have to estimate $R_{\Gamma_0}(D(\cdot, \Gamma_*))$ rather than $R_{\Gamma_0}(D(\cdot, \Gamma_0))$, and this is achieved by $n^{-1} \sum_1^n D(S(x_\nu, \vartheta^{(n)}(\underline{x})), \Gamma^{(n)}(\underline{x}))$, not by $R_{\Gamma^{(n)}(\underline{x})}(D(\cdot, \Gamma^{(n)}(\underline{x})))$.

Even if $\Gamma^{(n)}$, $n \in \mathbb{N}$, converges to the true Γ_0, the use of the latter estimator is justified only if $\Gamma \to R_\Gamma(D(\cdot, \Gamma))$ is weakly continuous. In our examples, $R_\Gamma(D(\cdot, \Gamma))$ is proportional to $A_{k,\ell}(\Gamma)$ with a factor depending continuously on ϑ, and

$$A_{k,\ell}(\Gamma) := \frac{1}{\Gamma(k)} \int_0^\infty s^{k+\ell} \frac{\left(\int_0^\infty \eta^{k+1} \exp[-\eta s] \Gamma(d\eta)\right)^2}{\int_0^\infty \eta^k \exp[-\eta s] \Gamma(d\eta)} \, ds.$$

Lemma L.30 shows that $A_{k,\ell}$ is continuous at Γ_0 for any $\Gamma_0 | \mathcal{C}$ if $\ell = 1$. It is not continuous in general for $\ell \neq 1$. Hence the use of the e.s. $R_{\Gamma^{(n)}}(D(\cdot, \Gamma^{(n)}))$ is justified by Lemma L.30 in Example 1, but not in Example 2 (where $\ell = 0$). In spite of this, the numerical results based on this estimator are satisfactory.

The results concerning the estimates $\hat{\vartheta}^{(n)}(\underline{x}, \vartheta^{(n)}(\underline{x}), \Gamma^{(n)}(\underline{x}))$ are listed under **improved A_0**.

Since the estimation procedure for Γ is rather crude, it seemed commendable to assess the possible gains of more refined estimation procedures. For this purpose, we investigated the performance of the estimates $\hat{\vartheta}^{(n)}(\underline{x}, \vartheta^{(n)}(\underline{x}), \Gamma_0)$, based on the true Γ_0. The results for these estimates are listed under **fictitious A**. It appears that not much can be gained from more refined estimation procedures for Γ.

Moreover, we applied in Examples 1 and 3 improvement procedure (9.16) with only 3 supporting points. The results for this case are listed under **improved B**. The increase of the as. mean deviation resulting from this simplification is surprisingly small.

For Examples 1 and 2, we also present results on improved e.s. based on (see (9.14), (9.29)) $N(x,\vartheta,\alpha) = -S^\bullet(x,\vartheta)/(\alpha + S(x,\vartheta))$.

Written explicitly, the improved e.s. (9.16) is now

$$\tilde{\vartheta}^{(n)}(\underline{x}) := \vartheta^{(n)}(\underline{x}) - \frac{\sum_1^n \frac{S^\bullet(x_\nu, \vartheta^{(n)}(\underline{x}))}{\alpha^{(n)}(\underline{x}) + S(x_\nu, \vartheta^{(n)}(\underline{x}))}}{C(\vartheta^{(n)}(\underline{x})) \sum_1^n \left[\frac{\dot{B}(S(x_\nu, \vartheta^{(n)}(\underline{x})))}{\alpha^{(n)}(\underline{x}) + S(x_\nu, \vartheta^{(n)}(\underline{x}))} - \frac{B(S(x_\nu, \vartheta^{(n)}(\underline{x})))}{(\alpha^{(n)}(\underline{x}) + S(x_\nu, \vartheta^{(n)}(\underline{x})))^2}\right]} \qquad (9.41)$$

with $\alpha^{(n)}(\underline{x}) := \hat{\alpha}^{(n)}(S(x_1, \vartheta^{(n)}(\underline{x})), \ldots, S(x_n, \vartheta^{(n)}(\underline{x})))$.

The pertaining results are listed under **improved C**. The theoretical value of the mean deviation is $\sqrt{2/\pi}\sigma(\vartheta_0, \Gamma_0, \alpha_*)$, where α_* is the value minimizing $\alpha \to \sigma(\vartheta_0, \Gamma_0, \alpha)$ (which is α_0 for $\Gamma_0 = \Gamma_{\alpha_0, \beta_0}$).

*   *

*

The following examples are taken from literature, where they served as illustrations for the application of estimating equations, approached with unknown rather than random nuisance parameters in mind. Since the relevance of results on random nuisance parameters for the case of unknown nuisance parameters is yet under discussion, we confine ourselves to the following statement:

(i) If the estimating equations suggested for these examples are applied in the case of random nuisance parameters, the resulting e.s. are as. inefficient.

(ii) The results of our simulation experiments confirm the promises of as. theory to a large extent: The reduction of the standard deviation by the improvement procedure turns out to be about as large as promised by as. theory. Theoretically, the bias should be 0 for all estimators. It turns out that it is negligible in all cases. In particular: The improvement procedures do not enlarge the bias of the initial estimator.

(iii) The numerical accuracy is in no close connection with the fulfilment of the regularity conditions.

(iv) The simulation experiments are not a precise model of what one would do in real applications. There, one would start with a preliminary choice of the supporting points, and one would use the estimate $\Gamma^{(n)}(\underline{x})$ to rectify this choice (i.e. to insert additional supporting points where the density is high, where it shows a second mode, etc.). Such a procedure will certainly lead to more satisfying results. The fact that it is too complex to be described by a mathematical model should not be an argument against its use if it works satisfactorily.

Example 1

Let $P_{\vartheta,\xi}|\mathbb{B}^2$, $\vartheta, \xi \in \mathbb{R}_+$, denote the p–measure with λ^2–density

$$p((x_1, x_2), \vartheta, \xi) = \vartheta\xi^2 \exp[-\xi(\vartheta x_1 + x_2)], \quad (x_1, x_2) \in \mathbb{R}_+^2. \tag{E.1.1}$$

With the transformed parameter $\eta = \vartheta^{1/2}\xi$ this density can be rewritten in the form of (9.1), with dominating measure $\mu = \lambda^2$,

$$S((x_1, x_2), \vartheta) = \vartheta^{-1/2}(\vartheta x_1 + x_2)$$

and

$$A(\eta) = \eta^2, \quad \text{hence } R_\eta = \Gamma_{\eta,2} \text{ and } q_0(s) = s.$$

We have

$$S^\bullet((x_1, x_2), \vartheta) = \frac{\vartheta^{-3/2}}{2}(\vartheta x_1 - x_2),$$
$$C(\vartheta) = \vartheta^{-2}, \quad \hat{B}(s) = s/4, \quad B(s) = s^2/12.$$

A transformation generating the family (E.1.1) is $(x_1, x_2) \to (\lambda x_1, \lambda x_2)$, $\lambda > 0$, which leads to $S((\lambda x_1, \lambda x_2), \vartheta) = \lambda S((x_1, x_2), \vartheta)$.

Now we shall show that conditions (9.21) and (9.23) are fulfilled for all p–measures $\Gamma_*|\mathbb{B}_+$ if

$$\int_1^\infty (\log \eta)^2 \Gamma_0(d\eta) < \infty \quad \text{and} \quad \int_0^1 \eta^{-2} \Gamma_0(d\eta) < \infty. \tag{E.1.2}$$

If $f_k(x_1, x_2, \vartheta)$ is a homogeneous polynomial of degree k in x_1, x_2 (with coefficients depending continuously on ϑ) we have for an appropriate neighborhood $V \ni \vartheta_0$ with a generic constant c

$$\sup_{\vartheta \in V} |f_k(x_1, x_2, \vartheta)| \leq c(x_1^k + x_2^k)$$

and

$$P_{\vartheta_0, \cdot}^{S(\cdot, \vartheta_0) = s}(x_1^k + x_2^k) \leq cs^k.$$

In the following we use this for $S^{\bullet\bullet}((x_1, x_2), \vartheta)$, a homogeneous polynomial of degree 1, and $S^{\bullet}((x_1, x_2), \vartheta)^2$, a homogeneous polynomial of degree 2.

Moreover, $\vartheta \in V_0 := \left(\frac{\vartheta_0}{4}, 4\vartheta_0\right)$ implies

$$S((x_1, x_2), \vartheta) \geq \frac{1}{2} S((x_1, x_2), \vartheta_0).$$

Since $s \mapsto |H_0(s, \Gamma)|$ is nonincreasing (by Lemma L.22), we obtain

$$P_{\vartheta_0, \Gamma_0}\left(\sup_{\vartheta \in V_0} \sup_{\Gamma \in U} |S^{\bullet\bullet}((x_1, x_2), \vartheta) H_0(S((x_1, x_2), \vartheta), \Gamma)|\right)$$

$$\leq P_{\vartheta_0, \Gamma_0}\left(\sup_{\vartheta \in V_0} |S^{\bullet\bullet}((x_1, x_2), \vartheta)| \sup_{\Gamma \in U} |H_0\left(\frac{1}{2} S((x_1, x_2), \vartheta_0), \Gamma\right)|\right)$$

$$\leq c \int s \sup_{\Gamma \in U} |H_0\left(\frac{1}{2} s, \Gamma\right)| R_{\Gamma_0}(ds).$$

Lemma L.23 implies for every $\Gamma_* \in \mathcal{G}$ the existence of a vague neighborhood $U \ni \Gamma_*$ and a generic constant c such that

$$\sup_{\Gamma \in U} |H_0(s, \Gamma)| \leq \begin{cases} c \frac{1+|\log s|}{s} & s \in (0, 1] \\ c & s > 1. \end{cases}$$

We obtain

$$\int s \sup_{\Gamma \in U} |H_0\left(\frac{1}{2} s, \Gamma\right)| R_{\Gamma_0}(ds)$$

$$\leq c \int_0^2 s \frac{1+|\log(s/2)|}{s/2} R_{\Gamma_0}(ds) + c \int_2^\infty s R_{\Gamma_0}(ds) < \infty$$

if

$$\int_1^\infty \log \eta \, \Gamma_0(d\eta) < \infty \quad \text{and} \quad \int_0^1 \eta^{-1} \Gamma_0(d\eta) < \infty.$$

Finally, since $H_2(\cdot, \Gamma)$ is nonincreasing too (see Lemma L.22), we obtain

$$P_{\vartheta_0, \Gamma_0}\left(\sup_{\vartheta \in V_0} \sup_{\Gamma \in U} S^{\bullet}((x_1, x_2), \vartheta)^2 H_2(S((x_1, x_2), \vartheta), \Gamma)\right)$$

$$\leq P_{\vartheta_0, \Gamma_0}\left(\sup_{\vartheta \in V_0} S^{\bullet}((x_1, x_2), \vartheta)^2 \sup_{\Gamma \in U} H_2\left(\frac{1}{2} S((x_1, x_2), \vartheta_0), \Gamma\right)\right)$$

$$\leq c \int_0^\infty s^2 \sup_{\Gamma \in U} H_2\left(\frac{1}{2} s, \Gamma\right) R_{\Gamma_0}(ds).$$

As a consequence of Lemma L.23, there exists a vague neighborhood $U \ni \Gamma_*$ such that
$$\sup_{\Gamma \in U} H_2(s, \Gamma) \leq \begin{cases} c \frac{1+(\log s)^2}{s^2} & s \in (0, 1] \\ c & s > 1. \end{cases}$$

Hence we obtain
$$\int_0^\infty s^2 \sup_{\Gamma \in U} H_2\left(\frac{1}{2}s, \Gamma\right) R_{\Gamma_0}(ds)$$
$$\leq 4c \int_0^2 \left(1 + (\log s)^2\right) R_{\Gamma_0}(ds) + c \int_2^\infty s^2 R_{\Gamma_0}(ds) < \infty$$

if
$$\int_1^\infty (\log \eta)^2 \Gamma_0(d\eta) < \infty \quad \text{and} \quad \int_0^1 \eta^{-2} \Gamma_0(d\eta) < \infty.$$

The as. variance bound for e.s. for ϑ is (see (9.8'))
$$\sigma_0^2(\vartheta, \Gamma) = \vartheta^2 \hat{\sigma}_0^2(\Gamma), \tag{E.1.3}$$

with
$$\hat{\sigma}_0^2(\Gamma) := 12 \left(\int_0^\infty s^3 \left[\int_0^\infty \eta^3 \exp[-\eta s] \Gamma(d\eta) \right]^2 \left(\int_0^\infty \eta^2 \exp[-\eta s] \Gamma(d\eta) \right)^{-1} ds \right)^{-1}.$$

Observe that $\hat{\sigma}_0^2(\Gamma_a) = \hat{\sigma}_0^2(\Gamma)$ if $\Gamma_a = \Gamma * (\eta \to a\eta)$, $a > 0$.

Written in a slightly different way, $\sigma_0^{-2}(\vartheta, \Gamma)$ occurs in Lindsay (1985, p. 920), without proof, as "information in the presence of mixing".

For Γ being a Gamma distribution we obtain
$$\hat{\sigma}_0^2(\Gamma_{\alpha, \beta}) = 2 \frac{\beta + 3}{\beta + 2}. \tag{E.1.4}$$

$$* \qquad * \qquad *$$

The as. variance of the improved e.s. based on $N(\cdot, \vartheta, \alpha)$ given by (9.29) is (see (9.32))
$$\sigma^2(\vartheta, \Gamma, \alpha) = \vartheta^2 \hat{\sigma}^2(\Gamma, \alpha), \tag{E.1.5}$$

with

$$\hat{\sigma}^2(\Gamma,\alpha) = \frac{1}{3}\int_0^\infty \frac{s^2}{(\alpha+s)^2} R_\Gamma(ds) \Big/ \Big(\int_0^\infty \frac{s}{(\alpha+s)^2}\Big(\frac{\alpha}{2}+\frac{s}{3}\Big) R_\Gamma(ds)\Big)^2.$$

Of course (see (9.37))
$$\hat{\sigma}^2(\Gamma_{\alpha,\beta},\alpha) = \hat{\sigma}_0^2(\Gamma_{\alpha,\beta}).$$

It is easy to see that conditions (9.35), (9.36), (9.38) and (9.39) are fulfilled provided $\int_0^1 \eta^{-2}\Gamma_0(d\eta) < \infty$.

* * *

Lindsay (1982, p. 505, Example 1) suggests (for a formally more general model) the estimating function

$$N((x_1,x_2),\vartheta) = \vartheta x_1 - x_2,$$

and in (1985, p. 914/5, Example A) the estimating function

$$N((x_1,x_2),\vartheta) = (\vartheta x_1 - x_2)/(\vartheta x_1 + x_2). \tag{E.1.6}$$

The idea to use the estimating function $(\vartheta x_1 - x_2)/(\alpha + (\vartheta x_1 + x_2))$ corresponding to $N(\cdot,\vartheta,\alpha)$ given by (9.29), occurs in Lindsay (1985, Section 5).

According to Kumon and Amari (1984, p. 457, Example 5) the estimating function (E.1.6) is optimal in their class C_2. The as. variance of the resulting e.s. is $3\vartheta^2$, an amount surpassing the as. variance bound given by (E.1.3).

That $\hat{\sigma}_0^2(\Gamma) < 3$ for all $\Gamma | I\!B_+$ can be seen as follows.

$$4 = \Big(\int_0^\infty\int_0^\infty \eta^3 s^2 e^{-\eta s}\Gamma(d\eta)ds\Big)^2$$

$$= \Big(\int_0^\infty s \frac{\int_0^\infty \eta^3 s e^{-\eta s}\Gamma(d\eta)}{\int_0^\infty \eta^2 s e^{-\eta s}\Gamma(d\eta)} \int_0^\infty \eta^2 s e^{-\eta s}\Gamma(d\eta)ds\Big)^2$$

$$\leq \int_0^\infty s^2 \frac{\big(\int_0^\infty \eta^3 s e^{-\eta s}\Gamma(d\eta)\big)^2}{\int_0^\infty \eta^2 s e^{-\eta s}\Gamma(d\eta)} ds = 12/\hat{\sigma}_0^2(\Gamma)$$

with strict inequality unless

$$s\frac{\int_0^\infty \eta^3 e^{-\eta s}\Gamma(d\eta)}{\int_0^\infty \eta^2 e^{-\eta s}\Gamma(d\eta)} = c(\Gamma) \quad \text{for } s > 0.$$

From this, we obtain

$$\int_0^\infty\int_0^\infty \eta^3 s^{k+1} e^{-\eta s} ds\Gamma(d\eta) = c(\Gamma)\int_0^\infty\int_0^\infty \eta^2 s^k e^{-\eta s} ds\Gamma(d\eta).$$

Applied for $k = 0$ and $k = 1$ we obtain a contradiction.

The as. variance bound (E.1.3) was given in Pfanzagl (1987, p. 245/6, Example 1), where also a part of the following numerical results was published in graphical form.

Van der Vaart (1988) treats this model as Example 5.7.3. Using modified kernel estimators for p_0'/p_0 no regularity conditions on Γ are required for the validity of the improvement procedure.

As a preliminary estimator we use

$$\vartheta^{(n)}\big((x_{1\nu}, x_{2\nu})_{\nu=1,\dots,n}\big) = \text{med}\{x_{2\nu}/x_{1\nu} : \nu = 1,\dots,n\},$$

which is as. normal with variance $4\vartheta^2$.

Sample size $n = 101$
Estimand $\vartheta = 1$
Mixing distribution $\Gamma\{1\} = 1$
Mean deviation, theor. 1.206
Simulations $N = 10\,000$

estimator	bias	mean deviation theor.	empir.	e.m.
preliminary	0.040	1.414	1.424	0.023
improved A_0	0.013	1	1.018	0.016
fictitious A_0	0.022	1	1.008	0.016

Sample size $n = 101$
Estimand $\vartheta = 1$
Mixing distribution $\Gamma = \Gamma_{(1,3)}$
Mean deviation, theor. 1.236
Simulations $N = 10\,000$

estimator	bias	mean deviation		
		theor.	empir.	e.m.
preliminary	0.038	1.291	1.302	0.021
improved A_0	0.007	1	1.067	0.017
fictitious A_0	0.004	1	1.003	0.016

Sample size $n = 101$
Estimand $\vartheta = 1$
Mixing distribution $\Gamma = \Gamma_{(1,5)}$
Mean deviation, theor. 1.206
Simulations $N = 10\,000$

estimator	bias	mean deviation				coverage
		theor.	empir.	e.m.	sample	
preliminary	0.046	1.323	1.356	0.029	1.355	0.897
improved A	0.010	1.001	1.010	0.020	·	·
improved B	0.014	1.007	1.014	0.020	1.028	0.891
improved C	0.010	1	1.031	0.020	·	·
ficticious C	0.009	1	1.010	0.020	·	·

A dot in the tables indicates that this number has not been computed.

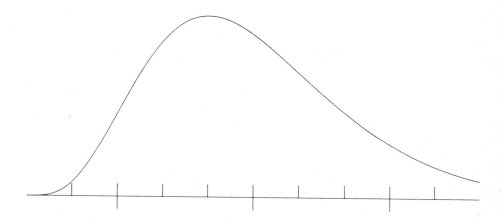

Fig. 1. Shows $\Gamma_{(1,5)}$ with the 9 supporting points used for A, and the 3 supporting points used for B.

The following simulation experiment shows how improvement procedure C works if the mixing distribution is not a Gamma distribution. In this case, the as. variance is not the minimal one, but — in spite of a "wrong" model — the mean deviation exceeds that of the as. efficient estimator only slightly, and the bias of the improved estimator remains small.

Sample size $n = 101$
Estimand $\vartheta = 1$
Mixing distribution $\Gamma = \frac{7}{20}\Gamma_{(20,\frac{11}{2})} + \frac{13}{20}\Gamma_{(\frac{3}{2},\frac{11}{2})}$
Mean deviation, theor. 1.265
Simulations $N = 10\,000$

| estimator | bias | mean deviation | | | sample | coverage |
		theor.	empir.	e.m.		
preliminary	0.050	1.261	1.288	0.027	1.294	0.902
improved A	0.022	1.002	1.021	0.021	.	.
improved B	0.011	1.033	1.049	0.021	1.059	0.892
improved C	0.024	1.035	1.047	0.021	.	.

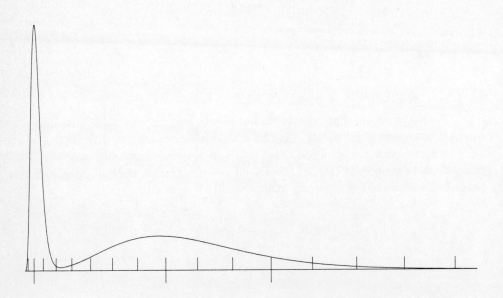

Fig. 2.   Shows $\frac{7}{20}\Gamma_{(20,\frac{11}{2})} + \frac{13}{20}\Gamma_{(\frac{3}{2},\frac{11}{2})}$ with the 16 supporting points used for A, and the 3 supporting points used for B.

Example 2

Let $P_{\vartheta,\sigma}|I\!B^q$, $\vartheta \in I\!R$, $\sigma \in I\!R_+$ denote the p-measure with λ^q-density

$$p((x_1,\ldots,x_q),\vartheta,\sigma) = (2\pi\sigma^2)^{-\frac{q}{2}} \exp[-\frac{1}{2\sigma^2}\sum_{i=1}^{q}(x_i - \vartheta)^2], \quad (E.2.1)$$

$(x_1,\ldots,x_q) \in I\!R^q$.

With the transformed parameter $\eta = \sigma^{-2}$ this can be rewritten in the form of (9.1) with dominating measure $\mu = \lambda^q$,

$$S((x_1,\ldots,x_q),\vartheta) = \frac{1}{2}\sum_{1}^{q}(x_i - \vartheta)^2$$

and

$$A(\eta) = (\eta/2\pi)^{q/2}, \text{ hence } R_\eta = \Gamma_{\eta,\,q/2} \text{ and } q_0(s) = \frac{s^{\frac{q}{2}-1}}{\Gamma(\frac{q}{2})}(2\pi)^{\frac{q}{2}}.$$

We have

$$S^{\bullet}((x_1,\ldots,x_q),\vartheta) = -\sum_{1}^{q}(x_i - \vartheta),$$

$$C(\vartheta) = 1, \quad \hat{B}(s) = q, \quad B(s) = 2s.$$

A transformation generating the family (E.2.1) is

$$(x_1,\ldots,x_q) \to (\sqrt{\lambda}x_1 + \vartheta(1-\sqrt{\lambda}),\ldots,\sqrt{\lambda}x_q + \vartheta(1-\sqrt{\lambda})), \quad \lambda > 0,$$

which transforms $S((x_1,\ldots,x_q),\vartheta)$ into $\lambda S((x_1,\ldots,x_q),\vartheta)$.

Now we shall show that conditions (9.21) and (9.23) are fulfilled for all p-measures $\Gamma_*|I\!B_+$ if

$$q > 3, \quad \int_1^\infty \eta(\log\eta)^2 \Gamma_0(d\eta) < \infty \quad \text{and} \quad \int_0^1 \eta^{-1}\Gamma_0(d\eta) < \infty. \quad (E.2.2)$$

We use

$$S^{\bullet\bullet}((x_1,\ldots,x_q),\vartheta) = q$$

and
$$S^{\bullet}((x_1,\ldots,x_q),\vartheta)^2 \leq 2qS((x_1,\ldots,x_q),\vartheta).$$

Hence we have to establish the existence of neighborhoods $V \ni \vartheta_0$ and $U \ni \Gamma_*$ such that

$$\int \sup_{\vartheta \in V} \sup_{\Gamma \in U} |H_0(S((x_1,\ldots,x_q),\vartheta),\Gamma)| P_{\vartheta_0,\Gamma_0}(d(x_1,\ldots,x_q)) < \infty$$

and

$$\int \sup_{\vartheta \in V} \sup_{\Gamma \in U} S((x_1,\ldots,x_q),\vartheta) H_2(S((x_1,\ldots,x_q),\vartheta),\Gamma) P_{\vartheta_0,\Gamma_0}(d(x_1,\ldots,x_q))$$
$$< \infty.$$

Lemma L.23 implies for every $\Gamma_* \in \mathcal{G}$ the existence of a vague neighborhood $U \ni \Gamma_*$ and a generic constant c such that

$$\sup_{\Gamma \in U} |H_0(s,\Gamma)| \leq \begin{cases} c\frac{1+|\log s|}{s} & s \in (0,1] \\ c & s > 1 \end{cases}$$

and

$$\sup_{\Gamma \in U} H_2(s,\Gamma) \leq \begin{cases} c\frac{1+(\log s)^2}{s^2} & s \in (0,1] \\ c & s > 1. \end{cases}$$

For any $\vartheta \in \mathbb{R}$,

$$\tilde{S}(x_1,\ldots,x_q) := \frac{1}{2}\sum_{1}^{q}(x_i - \overline{x}_q)^2 \leq \frac{1}{2}\sum_{1}^{q}(x_i - \vartheta)^2$$
$$= S((x_1,\ldots,x_q),\vartheta).$$

Since $s \mapsto |H_0(s,\Gamma)|$ is nonincreasing and since $P_{\vartheta_0,\Gamma_0} * \tilde{S}$ has density $h_{q-1}(\cdot,\Gamma_0)$, where

$$h_q(s,\Gamma_0) := \frac{1}{\Gamma(\frac{q}{2})} \int_0^\infty \eta^{\frac{q}{2}} s^{\frac{q-2}{2}} e^{-\eta s} \Gamma_0(d\eta),$$

we obtain

$$\int \sup_{\vartheta \in V} \sup_{\Gamma \in U} |H_0(S((x_1,\ldots,x_q),\vartheta),\Gamma)| P_{\vartheta_0,\Gamma_0}(d(x_1,\ldots,x_q))$$
$$\leq c \int_0^1 \frac{1+|\log s|}{s} h_{q-1}(s,\Gamma_0) ds + c \int_1^\infty h_{q-1}(s,\Gamma_0) ds < \infty$$

if $q > 3$ and $\int_1^\infty \eta \log \eta \Gamma(d\eta) < \infty$.

We have $\tilde{S}(x_1, \ldots, x_q) \leq S((x_1, \ldots, x_q), \vartheta)$. Hence $S((x_1, \ldots, x_q), \vartheta) \leq 1$ implies for $\Gamma \in U$,

$$S((x_1, \ldots, x_q), \vartheta) H_2(S((x_1, \ldots, x_q), \vartheta), \Gamma)$$
$$\leq c \frac{1 + (\log S((x_1, \ldots, x_q), \vartheta))^2}{S((x_1, \ldots, x_q), \vartheta)}$$
$$\leq c \frac{1 + (\log \tilde{S}(x_1, \ldots, x_q))^2}{\tilde{S}(x_1, \ldots, x_q)}.$$

Let $V := (0, 2\vartheta_0)$. Then $\vartheta \in V$ implies

$$S((x_1, \ldots, x_q), \vartheta) \leq S((x_1, \ldots, x_q), \vartheta_0) + \frac{q}{2}\vartheta_0^2.$$

Hence $S((x_1, \ldots, x_q), \vartheta) \geq 1$ implies for $\vartheta \in V$, $\Gamma \in U$

$$S((x_1, \ldots, x_q), \vartheta) H_2(S((x_1, \ldots, x_q), \vartheta), \Gamma)$$
$$\leq cS((x_1, \ldots, x_q), \vartheta) \leq cS((x_1, \ldots, x_q), \vartheta_0) + \frac{q}{2}c\vartheta_0^2.$$

Therefore,

$$\int \sup_{\vartheta \in V} \sup_{\Gamma \in U} S((x_1, \ldots, x_q), \vartheta) H_2(S((x_1, \ldots, x_q), \vartheta), \Gamma) P_{\vartheta_0, \Gamma_0}(d(x_1, \ldots, x_q))$$
$$\leq c \int \left[\frac{1 + (\log \tilde{S}(x_1, \ldots, x_q))^2}{\tilde{S}(x_1, \ldots, x_q)} + S((x_1, \ldots, x_q), \vartheta_0) + \frac{q}{2}\vartheta_0^2 \right] P_{\vartheta_0, \Gamma_0}(d(x_1, \ldots, x_q))$$
$$= c \left(\int_0^\infty \frac{1 + (\log s)^2}{s} h_{q-1}(s, \Gamma_0) ds + \int_0^\infty s h_q(s, \Gamma_0) ds + \frac{q}{2}\vartheta_0^2 \right) < \infty$$

if $q > 3$, $\int_1^\infty \eta (\log \eta)^2 \Gamma(d\eta) < \infty$ and $\int_0^1 \eta^{-1} \Gamma(d\eta) < \infty$.

The as. variance bound for e.s. for ϑ is (see (9.8′), recall $C(\vartheta) = 1$)

$$\sigma_0^2(\vartheta, \Gamma; q) = \hat{\sigma}_0^2(\Gamma; q) = \frac{1}{2}\Gamma\left(\frac{q}{2}\right) \left(\int_0^\infty s^{\frac{q}{2}} \left[\int_0^\infty \eta^{\frac{q}{2}+1} \exp[-\eta s] \Gamma(d\eta) \right]^2 \quad \text{(E.2.3)}$$
$$\cdot \left(\int_0^\infty \eta^{\frac{q}{2}} \exp[-\eta s] \Gamma(d\eta) \right)^{-1} ds \right)^{-1}.$$

(We indicate the dependence of $\hat{\sigma}_0^2(\Gamma)$ on q for reasons which will become clear later.) Observe that $\hat{\sigma}_0^2(\Gamma_a; q) = a\hat{\sigma}_0^2(\Gamma; q)$ for $\Gamma_a := \Gamma * (\eta \to a\eta)$.

For Γ being a Gamma distribution, we obtain

$$\hat{\sigma}_0^2(\Gamma_{\alpha,\beta}; q) = \alpha \frac{2\beta + q + 2}{q\beta(2\beta + q)}. \qquad (\text{E.2.4})$$

The numerical results below refer to the case $q = 2$. They are in good agreement with the theoretical results, despite the fact that condition (E.2.2) requires $q > 3$. Even though $\hat{\sigma}_0^2(\Gamma; q)$ is not continuous in Γ, the improvement procedure using the estimate $\hat{\sigma}_0^2(\Gamma^{(n)}(\underline{x}); q)$ works quite well.

*   *

*

The as. variance of the improved e.s. based on $N(\cdot, \vartheta, \alpha)$ given by (9.29) is (see (9.32))

$$\hat{\sigma}^2(\Gamma, \alpha; q) = 2\int \frac{s}{(\alpha + s)^2} R_\Gamma(ds) / \left(\int \frac{q\alpha + (q-2)s}{(\alpha + s)^2} R_\Gamma(ds)\right)^2. \qquad (\text{E.2.5})$$

Of course (see (9.37))

$$\hat{\sigma}^2(\Gamma_{\alpha,\beta}, \alpha; q) = \hat{\sigma}_0^2(\Gamma_{\alpha,\beta}; q).$$

To meet (9.35), (9.36), (9.38) and (9.39), we need $q > 3$, $\int_1^\infty \eta \Gamma_0(d\eta) < \infty$ and $\int_0^\infty \eta^{-1} \Gamma_0(d\eta) < \infty$ which is slightly less than (E.2.2).

*   *

*

There is a large number of papers on this example (or variants of it), starting with Neyman and Scott (1948, Section 2, p. 3), Kalbfleisch and Sprott (1970, Section 6), Cox and Hinkley (1974, p. 147), Morton (1981, p. 232, Example 4), Lindsay (1982, p. 509, Example 2), Kumon and Amari (1984, p. 456/7, Example 4).

In these papers, the estimating function

$$N((x_1, \ldots, x_q), \vartheta) = \sum_{i=1}^{q}(x_i - \vartheta) / \sum_{i=1}^{q}(x_i - \vartheta)^2$$

is suggested. According to Kumon and Amari this estimating function is optimal in their class C_2. For $q > 2$, the resulting estimators have as.variance

$$\hat{\sigma}_1^2(\Gamma; q) := \frac{1}{q-2} / \int_0^\infty \eta \Gamma(d\eta). \tag{E.2.6}$$

That $\hat{\sigma}_0^2(\Gamma; q) < \hat{\sigma}_1^2(\Gamma; q)$ for all $q > 2$ and all nondegenerate p-measures $\Gamma | I\!B_+$ can be seen as follows.

Using Schwarz's inequality, we obtain

$$\left(\int_0^\infty \int_0^\infty \eta^{\frac{q}{2}+1} s^{\frac{q}{2}-1} e^{-\eta s} ds \Gamma(d\eta) \right)^2$$

$$= \left(\int_0^\infty s^{-1/2} \frac{\int_0^\infty \eta^{\frac{q}{2}+1} s^{\frac{q}{2}-\frac{1}{2}} e^{-\eta s} \Gamma(d\eta)}{\int_0^\infty \eta^{\frac{q}{2}} s^{\frac{q}{2}-1} e^{-\eta s} \Gamma(d\eta)} \int_0^\infty \eta^{\frac{q}{2}} s^{\frac{q}{2}-1} e^{-\eta s} \Gamma(d\eta) ds \right)^2$$

$$\leq \int_0^\infty s^{-1} \int_0^\infty \eta^{\frac{q}{2}} s^{\frac{q}{2}-1} e^{-\eta s} \Gamma(d\eta) ds$$

$$\cdot \int_0^\infty \left(\frac{\int_0^\infty \eta^{\frac{q}{2}+1} s^{\frac{q}{2}-\frac{1}{2}} e^{-\eta s} \Gamma(d\eta)}{\int_0^\infty \eta^{\frac{q}{2}} s^{\frac{q}{2}-1} e^{-\eta s} \Gamma(d\eta)} \right)^2 \int_0^\infty \eta^{\frac{q}{2}} s^{\frac{q}{2}-1} e^{-\eta s} \Gamma(d\eta) ds.$$

Since $\int_0^\infty \int_0^\infty \eta^m s^{m-2} e^{-\eta s} ds \Gamma(d\eta) = \Gamma(m-1) \int_0^\infty \eta \Gamma(d\eta)$, this implies $\hat{\sigma}_0^2(\Gamma; q) \leq \hat{\sigma}_1^2(\Gamma; q)$. The inequality is strict unless

$$s \frac{\int_0^\infty \eta^{\frac{q}{2}+1} e^{-\eta s} \Gamma(d\eta)}{\int_0^\infty \eta^{\frac{q}{2}} e^{-\eta s} \Gamma(d\eta)} = c(\Gamma) \quad \text{for } s > 0.$$

This is impossible; see Example 1.

The as. variance bound (E.2.3) was given in Pfanzagl (1987, p. 246/7, Example 2), where also the following numerical results were published in graphical form.

The natural preliminary estimator for ϑ is

$$\vartheta^{(n)}\big((x_{1\nu},\ldots,x_{q\nu})_{\nu=1,\ldots,n}\big) = (nq)^{-1} \sum_{\nu=1}^{n} \sum_{i=1}^{q} x_{i\nu} \qquad (\text{E.2.7})$$

which is as. normal with variance $q^{-1} \int_0^\infty \eta^{-1}\Gamma(d\eta)$. For $\Gamma = \Gamma_{\alpha,\beta}$ this variance is $\alpha/q(\beta-1)$.

Sample size $\qquad n = 100$
Estimand $\qquad \vartheta = 0$
Mixing distribution $\qquad \Gamma\{1\} = 1$
Mean deviation, theor. $\qquad 0.564$
Simulations $\qquad N = 10\,000$

estimator	bias	mean deviation theor.	mean deviation empir.	e.m.
preliminary	-0.004	1	1.005	0.016
improved A_0	-0.004	1	1.012	0.016
fictitious A_0	-0.004	1	1.005	0.016

Sample size $\qquad n = 100$
Estimand $\qquad \vartheta = 0$
Mixing distribution $\qquad \Gamma = \Gamma_{(1,\frac{5}{2})}$
Mean deviation, theor. $\qquad 0.405$
Simulations $\qquad N = 10\,000$

estimator	bias	mean deviation theor.	mean deviation empir.	e.m.
preliminary	-0.011	1.139	1.124	0.018
improved A_0	-0.009	1	1.022	0.016
fictitious A_0	-0.009	1	1.003	0.016
improved C	-0.007	1	1.014	0.019

In connection with this example we mention a result about the dependence of $\hat{\sigma}_0(\Gamma;q)$ on q which might be of interest in comparing the as. efficiency of samples with different "group–sizes" q. If we have n samples of group–size q each, we have, in fact, nq observations. If we standardize the error $\vartheta^{(n,q)} - \vartheta$ (with $\vartheta^{(n,q)}$ based on n samples $(x_1, \ldots, x_q)$) by $(nq)^{1/2}(\vartheta_q^{(n,q)} - \vartheta)$, then the as. variance bound is $q\hat{\sigma}_0^2(\Gamma;q)$ (see (E.2.3)). Smaller group sizes means (in an intuitive sense) that we admit more inhomogeneity in our sample of size nq. Hence the as. variance bound, $q\hat{\sigma}_0^2(\Gamma;q)$, should be a decreasing function of q. A repeated application of Schwarz's inequality shows that this is, in fact, true.

Example 3

Let $P_{\vartheta,\xi} = N_{(-\xi,1)} \times N_{(-\vartheta\xi,1)}$, $\vartheta \in \mathbb{R}_+$, $\xi \in \mathbb{R}$. The λ^2-density of $P_{\vartheta,\xi}$ is

$$p((x_1,x_2),\vartheta,\xi) = \frac{1}{2\pi} \exp[-\frac{1}{2}((x_1+\xi)^2 + (x_2+\vartheta\xi)^2)], \quad (x_1,x_2) \in \mathbb{R}^2. \quad \text{(E.3.1)}$$

With the transformed parameter $\eta = \xi(1+\vartheta^2)^{1/2}$ this can be rewritten in the form of (9.1), with respect to the dominating measure $\mu = N_{(0,1)}^2$,

$$S((x_1,x_2),\vartheta) = (x_1 + \vartheta x_2)(1+\vartheta^2)^{-1/2}$$

and

$$A(\eta) = \exp[-\frac{1}{2}\eta^2], \text{ hence } R_\eta = N_{(-\eta,1)} \text{ and } q_0(s) = \frac{1}{\sqrt{2\pi}} e^{-\frac{s^2}{2}}.$$

We have

$$S^\bullet((x_1,x_2),\vartheta) = -(\vartheta x_1 - x_2)(1+\vartheta^2)^{-3/2},$$

$$C(\vartheta) = (1+\vartheta^2)^{-2}, \quad \hat{B}(s) = -s, \quad B(s) = 1.$$

A transformation generating the family (E.3.1) is

$$(x_1, x_2) \to (x_1 + \lambda(1+\vartheta^2)^{-1/2}, \; x_2 + \lambda\vartheta(1+\vartheta^2)^{-1/2}), \quad \lambda \in \mathbb{R},$$

which transforms $S((x_1,x_2),\vartheta)$ into $S((x_1,x_2),\vartheta) + \lambda$.

Now we shall show that conditions (9.21) and (9.23) are fulfilled for all p-measures $\Gamma_*|\mathbb{B}$ if

$$\int \eta^4 \Gamma_0(d\eta) < \infty. \quad \text{(E.3.2)}$$

If $f_k((x_1,x_2),\vartheta)$ is a homogeneous polynomial in (x_1,x_2) of degree k (with coefficients depending continuously on ϑ), we may choose a neighborhood V of ϑ_0 such that — with a generic constant c —

$$\sup_{\vartheta \in V} |f_k((x_1,x_2),\vartheta)| \leq c(|x_1|^k + |x_2|^k).$$

Moreover, we use the bounds

$$P^{S(\cdot,\vartheta_0)=s}_{\vartheta_0,\cdot}(|x_1|^k + |x_2|^k) \leq c(1+|s|^k).$$

Lemma L.27 implies for every $\Gamma_* \in \mathcal{G}$ the existence of a vague neighborhood $U \ni \Gamma_*$ and a generic constant c such that, for all $s \in \mathbb{R}$,

$$\sup_{\Gamma \in U} |H_0(s,\Gamma)| \leq c(1+|s|),$$

$$\sup_{\Gamma \in U} H_2(s,\Gamma) \leq c(1+s^2).$$

For $k=1$ we obtain in particular, with a generic constant c,

$$\sup_{\vartheta \in V} |S((x_1,x_2),\vartheta)| \leq c(|x_1|+|x_2|),$$

$$\sup_{\vartheta \in V} |S^{\bullet\bullet}((x_1,x_2),\vartheta)| \leq c(|x_1|+|x_2|).$$

Hence

$$P_{\vartheta_0,\Gamma_0}\bigl(\sup_{\vartheta \in V} \sup_{\Gamma \in U} |S^{\bullet\bullet}((x_1,x_2),\vartheta) H_0(S((x_1,x_2),\vartheta),\Gamma)|\bigr)$$

$$\leq c P_{\vartheta_0,\Gamma_0}\bigl(\sup_{\vartheta \in V} |S^{\bullet\bullet}((x_1,x_2),\vartheta)|(1+\sup_{\vartheta \in V}|S((x_1,x_2),\vartheta)|)\bigr)$$

$$\leq c \int (x_1^2 + x_2^2) P_{\vartheta_0,\Gamma_0}(d(x_1,x_2))$$

$$\leq c \int (1+s^2) R_{\Gamma_0}(ds) < \infty \quad \text{if} \quad \int \eta^2 \Gamma_0(d\eta) < \infty.$$

Since $S^{\bullet}$ is a polynomial of degree 1,

$$\sup_{\vartheta \in V} \bigl(S^{\bullet}((x_1,x_2),\vartheta)\bigr)^2 \leq c(x_1^2 + x_2^2),$$

hence

$$P_{\vartheta_0,\Gamma_0}\bigl(\sup_{\vartheta \in V} \sup_{\Gamma \in U} S^{\bullet}((x_1,x_2),\vartheta)^2 H_2(S((x_1,x_2),\vartheta),\Gamma)\bigr)$$

$$\leq c P_{\vartheta_0,\Gamma_0}\bigl(\sup_{\vartheta \in V} S^{\bullet}((x_1,x_2),\vartheta)^2 (1+\sup_{\vartheta \in V} S((x_1,x_2),\vartheta)^2)\bigr)$$

$$\leq c \int (x_1^2 + x_2^2)(1+(x_1^2+x_2^2)) P_{\vartheta_0,\Gamma_0}(d(x_1,x_2))$$

$$\leq c \int (1+s^4) R_{\Gamma_0}(ds) < \infty \quad \text{if} \quad \int \eta^4 \Gamma_0(d\eta) < \infty.$$

The as. variance bound for e.s. for ϑ is (see (9.8))
$$\sigma_0^2(\vartheta, \Gamma) = (1+\vartheta^2)^2 \hat{\sigma}_0^2(\Gamma), \tag{E.3.3}$$
with
$$\hat{\sigma}_0^2(\Gamma) = \left(\frac{1}{\sqrt{2\pi}} \int_{-\infty}^{\infty} \frac{(\int_{-\infty}^{\infty} \eta \exp[-\frac{1}{2}(s+\eta)^2]\Gamma(d\eta))^2}{\int_{-\infty}^{\infty} \exp[-\frac{1}{2}(s+\eta)^2]\Gamma(d\eta)} ds \right)^{-1}.$$

For $\Gamma = N_{(\mu, \sigma^2)}$ this is
$$\hat{\sigma}_0^2(N_{(\mu,\sigma^2)}) = \left(\mu^2 + \sigma^4/(1+\sigma^2) \right)^{-1}. \tag{E.3.4}$$

Several authors offer an estimating equation as a solution to (variants of) this estimating problem, namely the one based on
$$N((x_1, x_2), \vartheta) := (\vartheta x_1 - x_2)(x_1 + \vartheta x_2) \tag{E.3.5}$$
(see Sprent (1969, p. 42), Morton (1981, p. 231, Example 3), Kumon and Amari (1984, p. 454, Example 2), Amari (1987a, p. 82) and Amari and Kumon (1988, p. 1050, Example 1)). The resulting e.s. has as. variance
$$\sigma_1^2(\vartheta, \Gamma) = (1+\vartheta^2)^2 \hat{\sigma}_1^2(\Gamma), \tag{E.3.6}$$
with
$$\hat{\sigma}_1^2(\Gamma) = [1 + \int \eta^2 \Gamma(d\eta)] / (\int \eta^2 \Gamma(d\eta))^2.$$

(In comparing with Kumon and Amari, recall the transformation $\eta = \xi(1+\vartheta^2)^{1/2}$.) According to Kumon and Amari, this e.s. is as. optimal in C_2. A comparison with (E.3.3) shows that (E.3.6) is not the minimal as. variance.

That $\hat{\sigma}_0^2(\Gamma) \leq \hat{\sigma}_1^2(\Gamma)$ for all Γ (with equality for $\Gamma = N_{(0,\sigma^2)}$) follows easily from Schwarz's inequality:

$$\left(\int \eta^2 \Gamma(d\eta) \right)^2 = \left(\frac{1}{\sqrt{2\pi}} \int_{-\infty}^{+\infty} s \int_{-\infty}^{\infty} \eta \exp\left[-\frac{1}{2}(s+\eta)^2\right]\Gamma(d\eta) ds \right)^2$$

$$\leq \frac{1}{\sqrt{2\pi}} \int_{-\infty}^{+\infty}\int_{-\infty}^{+\infty} s^2 \exp\left[-\frac{1}{2}(s+\eta)^2\right] ds \Gamma(d\eta) \cdot$$

$$\cdot \frac{1}{\sqrt{2\pi}} \int_{-\infty}^{\infty} \frac{(\int_{-\infty}^{\infty} \eta \exp[-\frac{1}{2}(s+\eta)^2]\Gamma(d\eta))^2}{\int_{-\infty}^{\infty} \exp[-\frac{1}{2}(s+\eta)^2]\Gamma(d\eta)} ds$$

$$= \int_{-\infty}^{\infty} (1+\eta^2)\Gamma(d\eta)/\hat{\sigma}_0^2(\Gamma)^2.$$

The solution of the estimating equation resulting from (E.3.5) was used as a preliminary estimator for ϑ. This solution can be written as

$$\vartheta^{(n)}(\underline{x}) = c^{(n)}(\underline{x}) + \left(1 + c^{(n)}(\underline{x})^2\right)^{1/2},$$

with

$$c^{(n)}(\underline{x}) = \sum_{1}^{n}(x_{2\nu}^2 - x_{1\nu}^2) / \left(2\sum_{1}^{n} x_{1\nu}x_{2\nu}\right).$$

For $\Gamma = N_{(\mu,\sigma^2)}$ we have

$$\hat{\sigma}_1^2(N_{(\mu,\sigma^2)}) = (1 + \mu^2 + \sigma^2)/(\mu^2 + \sigma^2)^2$$

(so that the preliminary e.s. is as. efficient for $\mu = 0$).

The comparatively large sample size 250 was chosen for one particular reason. With a smaller sample size, say 100, the sample–estimate for the standard deviation of the preliminary estimate becomes extremely large in a small number of cases (say 1%). This does not impair the coverage of the confidence procedure, but it has a noticeable influence on the average over the N standard deviations.

Sample size $n = 250$
Estimand $\vartheta = 1$
Mixing distribution $\Gamma = N_{(1,\frac{1}{4})}$
Mean deviation, theor. 1.557
Simulations $N = 10\,000$

| estimator | bias | mean deviation | | | | coverage |
		theor.	empir.	e.m.	sample	
preliminary	0.028	1.230	1.262	0.025	1.266	0.900
improved A_0	0.004	1	1.001	0.020	.	.
improved A	0.007	1.002	1.004	0.020	1.027	0.902
improved B	0.010	1.003	1.009	0.020	1.027	0.901

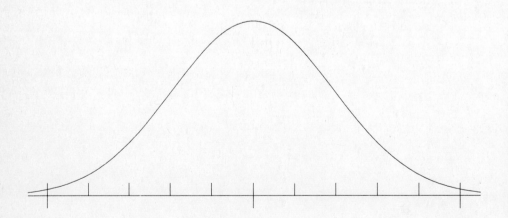

Fig. 3. Shows $N_{(1,\frac{1}{4})}$ with the 11 supporting points used for A, and the 3 supporting points used for B.

Sample size $n = 250$
Estimand $\vartheta = 1$
Mixing distribution $\Gamma = \frac{2}{3}N_{(\frac{3}{4},\frac{1}{16})} + \frac{1}{3}N_{(\frac{3}{2},\frac{1}{16})}$
Mean deviation, theor. 1.572
Simulations $N = 10\,000$

estimator	bias	mean deviation			sample	coverage
		theor.	empir.	e.m.		
preliminary	0.031	1.264	1.294	0.027	1.308	0.899
improved A_0	0.002	1	0.999	0.020	.	.
improved A	0.007	1.000	1.003	0.020	1.031	0.904
improved B	0.006	1.002	1.009	0.020	1.031	0.901

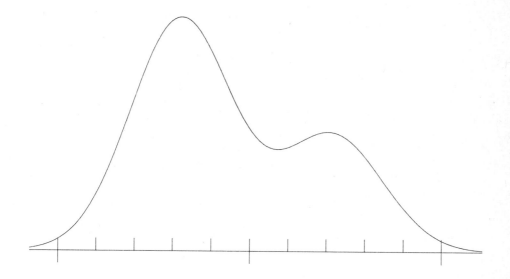

Fig. 4. Shows $\frac{2}{3}N_{(\frac{3}{4},\frac{1}{16})} + \frac{1}{3}N_{(\frac{3}{2},\frac{1}{16})}$ with the 11 supporting points used for A and the 3 supporting points used for B.

L. Auxiliary results

We repeatedly encounter the problem of replacing a parameter α by an estimator in expressions like $n^{-1/2}\sum_1^n f(x_\nu,\alpha)$ or $n^{-1}\sum_1^n f(x_\nu,\alpha)$. The following lemmas provide some technical tools for this purpose.

Let $(X,\mathcal{A})$ be a measurable space, $(Y,\mathcal{U})$ a Hausdorff space with countable base, $\nu|\mathcal{A}$ a measure, and $f: X\times Y \to \mathbb{R}$ a function with the following properties:

$$x \to f(x,y) \text{ is measurable for every } y \in Y, \qquad (\text{L.1})$$

$$y \to f(x,y) \text{ is continuous for every } x \in X. \qquad (\text{L.2})$$

To simplify our notations, we write for $B \subset Y$

$$\overline{f}(x,B) := \sup\{f(x,y) : y \in B\},$$
$$\underline{f}(x,B) := \inf\{f(x,y) : y \in B\}.$$

Lemma L.3. *For any set $B \subset Y$, the functions $\overline{f}(\cdot,B)$ and $\underline{f}(\cdot,B)$ are measurable if (L.1) and (L.2) hold.*

Proof. Since $(Y,\mathcal{U})$ has a countable base, there exists a countable set, say $B_0 \subset B$, which is dense in $(B, B \cap \mathcal{U})$.

Since B_0 is countable, $\overline{f}(x,B_0)$ is measurable. It remains to be shown that $\overline{f}(x,B_0) = \overline{f}(x,B)$ for $x \in X$.

Since $B_0 \subset B$, we have

$$\overline{f}(x,B_0) \leq \overline{f}(x,B).$$

For $r < \overline{f}(x,B)$, the set $\{y \in B : r < f(x,y)\}$ is non–empty, and open in $(B, B\cap\mathcal{U})$ by continuity of $y \to f(x,y)$. Hence it contains an element of B_0, say y_0, so that $r < f(x,y_0) \leq \overline{f}(x,B_0)$. □

Lemma L.4. *If (L.1) and (L.2) hold true, then $f: X\times Y \to \mathbb{R}$ is $\mathcal{A}\times\mathcal{B}$–measurable, where $\mathcal{B}$ is the Borel algebra of $(Y,\mathcal{U})$.*

Proof. We shall show that $S := \{(x,y) \in X\times Y : f(x,y) < r\} \in \mathcal{A}\times\mathcal{B}$ for every $r \in \mathbb{R}$.

Let $(x_0, y_0) \in S$ be arbitrary. Choose $r_0 < r$ such that $f(x_0, y_0) < r_0$. Since $y \to f(x,y)$ is continuous, the set $\{y \in Y : f(x_0, y) < r_0\}$ is an open set containing y_0.

Let $\mathcal{U}_0$ be a countable base of $\mathcal{U}$. There exists $U_0 \in \mathcal{U}_0$ such that $y_0 \in U_0 \subset \{y \in Y : f(x_0, y) < r_0\}$. Since $\overline{f}(x_0, U_0) \leq r_0 < r$, we have

$$(x_0, y_0) \in \bigcup_{U \in \mathcal{U}_0} \{x \in X : \overline{f}(x, U) < r\} \times U,$$

i.e.

$$S \subset \bigcup_{U \in \mathcal{U}_0} \{x \in X : \overline{f}(x, U) < r\} \times U,$$

since $(x_0, y_0) \in S$ was arbitrary.

Since the converse inclusion is obvious, we obtain

$$S = \bigcup_{U \in \mathcal{U}_0} \{x \in X : \overline{f}(x, U) < r\} \times U.$$

Since $\{x \in X : \overline{f}(x, U) < r\} \in \mathcal{A}$ by Lemma L.3, S is the countable union of $\mathcal{A} \times \mathcal{B}$-measurable sets and therefore measurable.

$\square$

Condition L.5. We shall say that $f : X \times Y \to \mathbb{R}$ fulfills condition L.5 for $y_0 \in Y$ and (a measure) $\nu | \mathcal{A}$ (L.5(y_0, ν) for short) if f fulfills (L.1) and (L.2) and if there exists an open neighborhood U of y_0 such that

$$\nu(\overline{f}(\cdot, U)) < \infty, \qquad \text{(L.5')}$$
$$\nu(\underline{f}(\cdot, U)) > -\infty. \qquad \text{(L.5'')}$$

Lemma L.6. *Let Y be an open convex subset of $\mathbb{R}^k$. If f fulfills L.5(y_0, ν), then*

$$f_0(x, y) := \int_0^1 f(x, (1-u)y_0 + uy) du$$

too fulfills L.5(y_0, ν).

Proof. (i) By Lemma L.4, $(x, y) \to f(x, y)$ is measurable. Hence $(x, u) \to f(x, (1-u)y_0 + uy)$ is measurable for each $y \in Y$, and $x \to f_0(x, y)$ is measurable by Fubini's Theorem.

(ii) Let $x \in X$ be fixed. Since $y \to f(x, y)$ is continuous on Y, it is uniformly continuous on any compact subset $Y_0 \subset Y$. Choose Y_0 as a convex neighborhood of y_0. Since $y_i \in Y_0$ implies $(1-u)y_0 + uy_i \in Y_0$, and

$$\|((1-u)y_0 + uy_1) - ((1-u)y_0 + uy_2)\| \leq \|y_1 - y_2\| \quad \text{for all } u \in [0, 1],$$

$y \to f_0(x,y)$ is uniformly continuous on Y_0. Since any $y \in Y$ can be included in such a set Y_0, $y \to f_0(x,y)$ is continuous on Y.

(iii) For any open convex set $U \ni y_0$, $y \in U$ implies $(1-u)y_0 + uy \in U$ for all $u \in [0,1]$. Hence

$$f(x, (1-u)y_0 + uy) \leq \overline{f}(x, U) \qquad \text{for all } u \in [0,1],$$

and $\overline{f}_0(x, U) \leq \overline{f}(x, U)$ follows. Since the set U in (L.5′) may be assumed convex w.l.g., (L.5′) for f_0 follows from (L.5′) for f. □

Lemma L.7. *Under assumption $L.5'(y_0, P)$ the following holds true.*

For every $\varepsilon > 0$ there exists an open set $U_\varepsilon \ni y_0$ such that

$$P(\overline{f}(\cdot, U_\varepsilon)) < P(f(\cdot, y_0)) + \varepsilon.$$

Proof. (i) Let U_n, $n \in \mathbb{N}$, be a nonincreasing local base at y_0. Then $\overline{f}(x, U_n) \downarrow f(x, y_0)$ for every $x \in X$.

For n sufficiently large, we have $U_n \subset U$ and therefore $\overline{f}(\cdot, U_n) \leq \overline{f}(\cdot, U)$. Hence Fatou's Lemma implies

$$P(\overline{f}(\cdot, U_n)) \downarrow P(f(\cdot, y_0)).$$

□

Corollary L.8. *Under assumption $L.5'(y_0, P)$ $[L.5''(y_0, P)]$, the function $y \to P(f(\cdot, y))$ is upper [lower] semi–continuous at y_0.*

In Gong and Samaniego (1981, p. 862, Lemma 2) the following assertion is proved under differentiability conditions.

Proposition L.9. *Assume $L.5(y_0, P)$. If $y^{(n)} : X^n \to Y$ fulfills $y^{(n)} = y_0 + o_p(n^0)$ (P^n), then*

$$n^{-1} \sum_1^n f(x_\nu, y^{(n)}(\underline{x})) = P(f(\cdot, y_0)) + o_p(n^0) \qquad (P^n).$$

Proof. By Lemma L.7, for every $\varepsilon > 0$ there exists an open $U_\varepsilon \ni y_0$ such that

$$P(\overline{f}(\cdot, U_\varepsilon)) < P(f(\cdot, y_0)) + \frac{\varepsilon}{2},$$
$$P(\underline{f}(\cdot, U_\varepsilon)) > P(f(\cdot, y_0)) - \frac{\varepsilon}{2}.$$

This implies

$$P^n\{|n^{-1}\sum_1^n (f(x_\nu, y^{(n)}(\underline{x})) - P(f(\cdot, y_0)))| > \varepsilon\}$$

$$\leq P^n\{y^{(n)} \notin U_\varepsilon\} + P^n\{n^{-1}\sum_1^n (\overline{f}(x_\nu, U_\varepsilon) - P(\overline{f}(\cdot, U_\varepsilon))) > \frac{\varepsilon}{2}\}$$

$$+ P^n\{n^{-1}\sum_1^n (\underline{f}(x_\nu, U_\varepsilon) - P(\underline{f}(\cdot, U_\varepsilon))) < -\frac{\varepsilon}{2}\}.$$

The assertion now follows from the law of large numbers.

$\square$

Proposition L.10. Let $\Theta \subset \mathbb{R}^m$, and $(A, \mathcal{U})$ a Hausdorff space with countable base. Assume that $h : X \times \Theta \times A \longrightarrow \mathbb{R}$ is differentiable in ϑ in a neighborhood of $\vartheta_0 \in \Theta^\circ$ (the interior of Θ), with partial derivatives

$$h^{(i)}(x, \vartheta_1, \ldots, \vartheta_m, \alpha) = \frac{\partial}{\partial \vartheta_i} h(x, \vartheta_1, \ldots, \vartheta_m, \alpha).$$

Assume that $h^{(i)}$ fulfills condition L.5$((\vartheta_0, \alpha_0), P)$ with $Y = \Theta \times A$. Let $\vartheta^{(n)} : X^n \to \Theta$ and $\alpha^{(n)} : X^n \to A$ fulfill

$$\vartheta^{(n)} = \vartheta_0 + O_p(n^{-1/2}) \quad (P^n),$$
$$\alpha^{(n)} = \alpha_0 + o_p(n^0) \quad (P^n).$$

Then

$$n^{-1/2}\sum_1^n \left(h(x_\nu, \vartheta^{(n)}(\underline{x}), \alpha^{(n)}(\underline{x})) - h(x_\nu, \vartheta_0, \alpha^{(n)}(\underline{x}))\right)$$

$$= \sum_1^m n^{1/2}(\vartheta_i^{(n)}(\underline{x}) - \vartheta_{0i}) P(h^{(i)}(\cdot, \vartheta_0, \alpha_0)) + o_p(n^0) \quad (P^n).$$

Corollary L.11.

$$n^{-1/2}\sum_1^n \left(h(x_\nu, \vartheta^{(n)}(\underline{x}), \alpha^{(n)}(\underline{x})) - h(x_\nu, \vartheta^{(n)}(\underline{x}), \alpha_0)\right)$$

$$= n^{-1/2}\sum_1^n \left(h(x_\nu, \vartheta_0, \alpha^{(n)}(\underline{x})) - h(x_\nu, \vartheta_0, \alpha_0)\right) + o_p(n^0) \quad (P^n).$$

For later use we remark that Proposition L.10 and the Corollary remain true if $n^{-1/2}\sum_{\nu=1}^{n}$ is replaced by $m_n^{-1/2}\sum_{\nu=1}^{m_n}$, provided m_n/n, $n \in \mathbb{N}$, is bounded away from 0.

Proof. We have

$$n^{-1/2}\sum_{1}^{n}\left(h(x_\nu, \vartheta^{(n)}(\underline{x}), \alpha^{(n)}(\underline{x})) - h(x_\nu, \vartheta_0, \alpha^{(n)}(\underline{x}))\right)$$

$$= \sum_{i=1}^{m} n^{1/2}(\vartheta_i^{(n)}(\underline{x}) - \vartheta_{0i}) n^{-1}\sum_{\nu=1}^{n}\int_0^1 h^{(i)}\left(x_\nu, (1-u)\vartheta_0 + u\vartheta^{(n)}(\underline{x}), \alpha^{(n)}(\underline{x})\right) du.$$

Since $(x, (\vartheta, \alpha)) \to \int_0^1 h^{(i)}(x, (1-u)\vartheta_0 + u\vartheta, \alpha) du$ fulfills L.5$((\vartheta_0, \alpha_0), P)$ by Lemma L.6, we obtain

$$n^{-1}\sum_{1}^{n}\int_0^1 h^{(i)}\left(x_\nu, (1-u)\vartheta_0 + u\vartheta^{(n)}(\underline{x}), \alpha^{(n)}(\underline{x})\right) du$$

$$= P\big(h^{(i)}(\cdot, \vartheta_0, \alpha_0)\big) + o_p(n^0) \qquad (P^n)$$

by Proposition L.9. Hence the assertion follows. $\square$

Lemma L.12. *Let $\Delta_m : X \times X^m \to \mathbb{R}$, $m \in \mathbb{N}$, be a sequence of measurable functions with the following properties.*

(i) $\int \Delta_m(x; x_1, \ldots, x_m) P(dx) = 0$ for all $(x_1, \ldots, x_m) \in X^m$,

(ii) $\int \Delta_m(x; x_1, \ldots, x_m)^2 P(dx) = o_p(m^0) \quad (P^m)$.

Then the following is true: For any sequence $m_n \in \mathbb{N}$, $n \in \mathbb{N}$, such that $n - m_n \to \infty$,

$$n^{-1/2}\sum_{\nu=1}^{m_n}\Delta_{n-m_n}(x_\nu; x_{m_n+1}, \ldots, x_n) = o_p(n^0) \quad (P^n).$$

Proof. We have

$$P^n\{(x_1, \ldots, x_n) \in X^n : |n^{-1/2}\sum_{1}^{m_n}\Delta_{n-m_n}(x_\nu; x_{m_n+1}, \ldots, x_n)| > \varepsilon\}$$

$$= \int P^{m_n}\{(x_1, \ldots, x_{m_n}) \in X^{m_n} : |n^{-1/2}\sum_{1}^{m_n}\Delta_{n-m_n}(x_\nu; x_{m_n+1}, \ldots, x_n)| > \varepsilon\}$$

$$P(dx_{m_n+1})\ldots P(dx_n).$$

Since
$$P^{m_n}\{(x_1,\ldots,x_{m_n}) \in X^{m_n} : |n^{-1/2} \sum_1^{m_n} \Delta_{n-m_n}(x_\nu; x_{m_n+1},\ldots,x_n)| > \varepsilon\}$$
$$\leq \varepsilon^{-2} \int \Delta_{n-m_n}(x; x_{m_n+1},\ldots,x_n)^2 P(dx) = o_p(n^0) \quad (P^{n-m_n})$$

the assertion follows by the Bounded Convergence Theorem. (Hint: use that the left-hand side is bounded.)

□

Lemma L.13. *Let $\Delta_n : X^{1+n} \to \mathbb{R}$, $n \in \mathbb{N}$, be a sequence of measurable functions, with $\Delta_n(x; x_1,\ldots,x_n)$ invariant under permutations of $(x_1,\ldots,x_n)$, which have the following properties.*

$$\int \Delta_n(x; x_1,\ldots,x_n) P(dx) = 0 \quad \text{for all } (x_1,\ldots,x_n) \in X^n, \tag{L.14}$$

$$\int \Delta_n(x; x_1,\ldots,x_n)^2 P(dx) P(dx_1) \ldots P(dx_n) = o(n^0) \tag{L.15}$$

$$\int \Delta_n(x; x_1,\ldots,x_n)^2 P(dx) P(dx_1) \ldots P(dx_n) \tag{L.16}$$
$$= \int \left(\int \Delta_n(x; x_1,\ldots,x_{n-1}, x_n) P(dx_n)\right)^2 P(dx) P(dx_1) \ldots P(dx_{n-1})$$
$$+ o(n^{-1}).$$

Then
$$n^{-1/2} \sum_1^n \Delta_{n-1}(x_\nu, x_{n\cdot\nu}) = o_p(n^0) \quad (P^n)$$

with $x_{n\cdot\nu} = (x_1,\ldots,x_{\nu-1}, x_{\nu+1},\ldots,x_n)$.

Proof. We have
$$P^n\{\underline{x} \in X^n : |n^{-1/2} \sum_{\nu=1}^n \Delta_{n-1}(x_\nu, x_{n\cdot\nu})| > \varepsilon\}$$
$$\leq \varepsilon^{-2} n^{-1} \sum_{\nu=1}^n \sum_{\mu=1}^n P^n\big(\Delta_{n-1}(x_\nu, x_{n\cdot\nu})\Delta_{n-1}(x_\mu, x_{n\cdot\mu})\big)$$
$$= \varepsilon^{-2} \big(P^n(\Delta_{n-1}(x_1, x_{n\cdot 1})^2)$$
$$+ (n-1)P^n\big(\Delta_{n-1}(x_1, x_{n\cdot 1})\Delta_{n-1}(x_2, x_{n\cdot 2})\big) = o(n^0),$$

since

$$P^n\big(\Delta_{n-1}(x_1, x_{n\cdot 1})\Delta_{n-1}(x_2, x_{n\cdot 2})\big)$$

$$= P^n\big((\Delta_{n-1}(x_1, x_{n\cdot 1}) - \int \Delta_{n-1}(x_1; y_2, x_{n\cdot 1,2})P(dy_2))$$

$$\cdot (\Delta_{n-1}(x_2, x_{n\cdot 2}) - \int \Delta_{n-1}(x_2; y_1, x_{n\cdot 1,2})P(dy_1))\big)$$

$$\leq P^n\big((\Delta_{n-1}(x_1, x_{n\cdot 1}) - \int \Delta_{n-1}(x_1; y_2, x_{n\cdot 1,2})P(dy_2))^2\big)$$

$$= o(n^{-1}) \quad \text{by (L.16)}.$$

$\square$

Lemma L.17. *Let $(X, \mathcal{A})$ be a measurable space, $P|\mathcal{A}^n$, $n \geq 2$, a permutation invariant p–measure, and $f : X^n \to \mathbb{R}$ a measurable function.*

Let Π_n denote the class of all permutations $\pi_n : X^n \to X^n$. For any $\underline{x} \in X^n$, let $f_*(\underline{x})$ denote the median of $\{f(\pi_n \underline{x}) : \pi_n \in \Pi_n\}$.

Then, for any $r > 0$,

$$P\{|f_*| > r\} \leq 2P\{|f| > r\}. \tag{L.18}$$

(Since the number of permutations is even, say $2m$, define the median of $(z_1, \ldots, z_{2m})$ by $z_{m:2m}$.)

Proof. For any function $h : X^n \to \mathbb{R}$, the function h_0 defined by $h_0(\underline{x}) := \frac{1}{n!}\sum_{\pi_n} h(\pi_n \underline{x})$ is in $P^{\mathcal{A}_0}(h)$, where $\mathcal{A}_0$ denotes the σ–algebra of all permutation invariant subsets of $\mathcal{A}^n$. Applied for $h = 1_{A_0} 1_{\{f \geq f_*\}}$ with $A_0 \in \mathcal{A}_0$ we obtain that $P^{\mathcal{A}_0}(h)$ contains the function

$$h_0 = 1_{A_0} \frac{1}{n!} \sum_{\pi_n} 1_{\{f \circ \pi_n \geq f_*\}} \geq \frac{1}{2} 1_{A_0}.$$

Hence

$$P\{\underline{x} \in A_0 : f(\underline{x}) \geq f_*(\underline{x})\} = \int h_0(\underline{x}) P(d\underline{x}) \geq \frac{1}{2} P(A_0).$$

Applied with $A_0 := \{f_* > r\}$ we obtain

$$P\{f \geq f_* > r\} \geq \frac{1}{2} P\{f_* > r\},$$

hence

$$P\{f_* > r\} \leq 2P\{f > r\}.$$

This proves one of the inequalities needed for (L.18). The other one follows with f replaced by $-f$.

$\square$

Let $(X, \mathcal{A})$ be a measurable space, $Y \subset \mathbb{R}^m$ and $h_n : X \times Y \to \mathbb{R}$ a measurable function. Assume that for any sequence $y_n = y_0 + O(n^{-1/2})$, with y_0 in the interior of Y, $h_n(\cdot, y_n) = o_p(n^0)$ (P).

The following lemma gives conditions under which the (nonstochastic) sequence y_n may be replaced by a stochastic one.

Lemma L.19. Let $(Z, \mathcal{C})$ be a measurable space, and $Q|\mathcal{C}$ a p–measure. If $y^{(n)} : Z \to Y$ fulfills $y^{(n)} = y_0 + O_p(n^{-1/2})$ (Q), then

$$(x, z) \to h_n(x, y^{(n)}(z)) = o_p(n^0) \quad (P \times Q).$$

Proof. For $\varepsilon > 0$ there exists c_ε such that

$$Q\{z \in Z : \|y^{(n)}(z) - y_0\| > c_\varepsilon n^{-1/2}\} < \varepsilon \quad \text{for all } n \in \mathbb{N}.$$

Define

$$\overline{y}^{(n)}(z) := \begin{cases} y^{(n)}(z) & \|y^{(n)}(z) - y_0\| \leq c_\varepsilon n^{-1/2}. \\ y_0 & > \end{cases}$$

We have

$$P \times Q\{(x, z) \in X \times Z : |h_n(x, y^{(n)}(z))| > \delta\}$$
$$\leq Q\{z \in Z : y^{(n)}(z) \neq \overline{y}^{(n)}(z)\}$$
$$+ P \times Q\{(x, z) \in X \times Z : |h_n(x, \overline{y}^{(n)}(z))| > \delta\}.$$

For every $z \in Z$ we have $\|\overline{y}^{(n)}(z) - y_0\| \leq c_\varepsilon n^{-1/2}$, hence

$$P\{x \in X : |h_n(x, \overline{y}^{(n)}(z))| > \delta\} = o(n^0)$$

for every $z \in Z$. Integration over z with respect to Q yields

$$P \times Q\{(x, z) \in X \times Z : |h_n(x, \overline{y}^{(n)}(z))| > \delta\} = o(n^0).$$

Hence the assertion follows.

$\square$

Discretization–Lemma L.20. Assume that $y^{(n)} : X \to Y$ fulfills $y^{(n)} = y_0 + O_p(n^{-1/2})$ (P).

For $r = (r_1, \ldots, r_m) \in \mathbb{R}^m$ let $d_n(r) := n^{-1/2}([n^{1/2}r_1], \ldots, [n^{1/2}r_m])$. Then the following holds true:

(i) $d_n(y^{(n)}) = y_0 + O_p(n^{-1/2})$ (P),

(ii) $(x \to h_n(x, d_n(y^{(n)}(x)))) = o_p(n^0)$ (P).

Proof. For $\varepsilon > 0$ there exists c_ε such that

$$P\{x \in X : \|y^{(n)}(x) - y_0\| > c_\varepsilon n^{-1/2}\} < \varepsilon \quad \text{for all } n \in \mathbb{N}.$$

For $n \in \mathbb{N}$, $k \in \mathbb{Z}^m$ let

$$y_{n,k} := d_n(y_0) + n^{-1/2} k.$$

By assumption,

$$P\{x \in X : |h_n(x, y_{n,k})| > \delta\} = o(n^0) \quad \text{for every } k \in \mathbb{Z}^m.$$

Therefore,

$$P\{x \in X : \sup_{\|k\| \leq c_\varepsilon + m^{1/2}} |h_n(x, y_{n,k})| > \delta\} = o(n^0).$$

If $\|y - y_0\| \leq c_\varepsilon n^{-1/2}$, then

$$d_n(y) \in \{y_{n,k} : k \in \mathbb{Z}^m, \|k\| \leq c_\varepsilon + m^{1/2}\}.$$

Hence $\|y^{(n)}(x) - y_0\| \leq c_\varepsilon n^{-1/2}$ implies

$$|h_n(x, d_n(y^{(n)}(x)))| \leq \sup_{\|k\| \leq c_\varepsilon + m^{1/2}} |h_n(x, y_{n,k})|,$$

so that

$$P\{x \in X : |h_n(x, d_n(y^{(n)}(x)))| > \delta\}$$
$$\leq P\{x \in X : \|y^{(n)}(x) - y_0\| > c_\varepsilon n^{-1/2}\}$$
$$+ P\{x \in X : \sup_{\|k\| \leq c_\varepsilon + m^{1/2}} |h_n(x, y_{n,k})| > \delta\}$$
$$\leq \varepsilon + o(n^0).$$

Hence (ii) follows. (i) is obvious. $\square$

Lemma L.21. Let $(X, \mathcal{U})$, $(Y, \mathcal{V})$ be topological spaces. Assume that $q : X \times Y \to \mathbb{R}$ has the following properties.

(i) $q(\cdot, y)$ is y-equicontinuous,

(ii) $q(x, \cdot) \in C_0(Y)$ for every $x \in X$.

Then
$$(x, M) \to \int q(x, y) M(dy)$$
is continuous with respect to $\mathcal{U} \times \mathcal{W}$ (with $\mathcal{W}$ denoting the topology of vague convergence in the class of all sub–probability measures on the Borel algebra of $\mathcal{V}$).

Proof. Let $\varepsilon > 0$ be arbitrary. For $x_0 \in X$ there exists a neighborhood $U_\varepsilon \in \mathcal{U}$ such that
$$|q(x, y) - q(x_0, y)| < \frac{\varepsilon}{2} \quad \text{for } x \in U_\varepsilon, \; y \in Y.$$

For any sub–probability measure M_0 there exists a neighborhood $W_\varepsilon \in \mathcal{W}$ such that
$$\left| \int q(x_0, y) M(dy) - \int q(x_0, y) M_0(dy) \right| < \frac{\varepsilon}{2} \quad \text{for } M \in W_\varepsilon.$$
Hence $x \in U_\varepsilon$, $M \in W_\varepsilon$ implies
$$\left| \int q(x, y) M(dy) - \int q(x_0, y) M_0(dy) \right| < \varepsilon.$$

□

Lemma L.22. Let $m : \mathbb{R} \to \mathbb{R}$ be a nondecreasing function and $\nu | \mathcal{B}$ a σ–finite measure.

(i) The set of all $s \in \mathbb{R}$ for which $\int e^{su} \nu(du)$ and $\int m(u) e^{su} \nu(du)$ exist, is an interval, say I (finite or infinite).

(ii) The function $s \to \int m(u) e^{su} \nu(du) / \int e^{su} \nu(du)$ is nondecreasing in the interior of I.

Proof. For (i) and the differentiability of the function in (ii) see Lehmann (1986, p. 57, Lemma 7, and p. 59, Theorem 9). Denoting by Q_s the p–measure with ν–density $u \to e^{su} / \int e^{sv} \nu(dv)$, the derivative can be written as
$$\int u m(u) Q_s(du) - \int m(u) Q_s(du) \int v Q_s(dv)$$
$$= \int \left(u - \int v Q_s(dv) \right) m(u) Q_s(du)$$
$$= \int \left(u - \int v Q_s(dv) \right) \left(m(u) - m\left(\int v Q_s(dv) \right) \right) Q_s(du) \geq 0.$$

□

Lemma L.23. Let $k > 0$ be fixed. Let $\Gamma_* | I\!B_+$ be a sub–probability measure with $\Gamma_*(I\!R_+) > 0$.

Then there exists a constant c and a vague neighborhood $U \ni \Gamma_*$ such that for all $\Gamma \in U$

$$\frac{\int_0^\infty \eta^{k+1} e^{-\eta s} \Gamma(d\eta)}{\int_0^\infty \eta^k e^{-\eta s} \Gamma(d\eta)} \leq \begin{cases} c \frac{1+|\log s|}{s} & s \in (0,1] \\ c & s \in (1, \infty) \end{cases}.$$

Proof. (i) Let $s \in (0,1]$. By Jensen's inequality,

$$\frac{\int_0^\infty \eta^{k+1} e^{-\eta s} \Gamma(d\eta)}{\int_0^\infty \eta^k e^{-\eta s} \Gamma(d\eta)} = \frac{2}{s} \cdot \frac{\int_0^\infty \log(e^{\eta s/2}) \eta^k e^{-\eta s} \Gamma(d\eta)}{\int_0^\infty \eta^k e^{-\eta s} \Gamma(d\eta)} \quad (L.24)$$

$$\leq \frac{2}{s} \log \frac{\int_0^\infty \eta^k e^{-\eta s/2} \Gamma(d\eta)}{\int_0^\infty \eta^k e^{-\eta s} \Gamma(d\eta)}.$$

For $s \in (0,1]$ we have

$$\log \int_0^\infty \eta^k e^{-\eta s} \Gamma(d\eta) \geq \log \int_0^\infty \eta^k e^{-\eta} \Gamma(d\eta) \quad (L.25)$$

$$> \log\left(\frac{1}{2} \int_0^\infty \eta^k e^{-\eta} \Gamma_*(d\eta)\right)$$

for all $\Gamma \in U$, where

$$U := \{\Gamma \in \mathcal{G}_* : \int_0^\infty \eta^k e^{-\eta} \Gamma(d\eta) > \frac{1}{2} \int_0^\infty \eta^k e^{-\eta} \Gamma_*(d\eta)\}.$$

Moreover, $\eta^k e^{-\eta s/2} \leq c_k s^{-k}$ for all $\eta > 0$, hence

$$\log \int_0^\infty \eta^k e^{-\eta s/2} \Gamma(d\eta) \leq \hat{c}_k (1 + |\log s|) \quad (L.26)$$

for any $\Gamma \in \mathcal{G}_*$.

The bound for $s \in (0, 1]$ follows from (L.24), (L.25) and (L.26).

(ii) Since
$$s \to \int_0^\infty \eta^{k+1} e^{-\eta s} \Gamma(d\eta) / \int_0^\infty \eta^k e^{-\eta s} \Gamma(d\eta)$$
is nonincreasing by Lemma L.22, we obtain from the bound in (i), applied for $s = 1$,
$$\frac{\int_0^\infty \eta^{k+1} e^{-\eta s} \Gamma(d\eta)}{\int_0^\infty \eta^k e^{-\eta s} \Gamma(d\eta)} \leq \frac{\int_0^\infty \eta^{k+1} e^{-\eta} \Gamma(d\eta)}{\int_0^\infty \eta^k e^{-\eta} \Gamma(d\eta)} \leq c \quad \text{for } \Gamma \in U, \ s > 1.$$

$\Box$

Lemma L.27. *Let $\Gamma_* | \mathbb{B}$ be a sub–probability measure with $\Gamma_*(\mathbb{R}) > 0$.*

Then there exists a constant c and a vague neighborhood $U \ni \Gamma_*$ such that for all $\Gamma \in U$, $s \in \mathbb{R}$
$$\frac{\int_{-\infty}^{+\infty} |\eta| e^{-(\eta-s)^2/2} \Gamma(d\eta)}{\int_{-\infty}^{+\infty} e^{-(\eta-s)^2/2} \Gamma(d\eta)} \leq c(1 + |s|).$$

Proof. It suffices to prove the assertion for Γ concentrated on $[0, \infty)$.

(i) We have
$$F_\Gamma(s) := \int_0^\infty \eta e^{-(\eta-s)^2/2} \Gamma(d\eta) / \int_0^\infty e^{-(\eta-s)^2/2} \Gamma(d\eta)$$
$$= \int_0^\infty \eta e^{\eta s - \eta^2/2} \Gamma(d\eta) / \int_0^\infty e^{\eta s - \eta^2/2} \Gamma(d\eta).$$

Since F_Γ is nondecreasing by Lemma L.22, we have $F_\Gamma(s) \leq F_\Gamma(1)$ for $s \leq 1$. Since $\Gamma \to F_\Gamma(1)$ is vaguely continuous, $U_0' := \{\Gamma \in \mathcal{G}_* : F_\Gamma(1) < 2 F_{\Gamma_*}(1)\}$ is open, and we have
$$F_\Gamma(s) \leq 2 F_{\Gamma_*}(1) \quad \text{for } \Gamma \in U_0', \ s \leq 1. \tag{L.28}$$

(ii) Assume now $s > 1$. We have

$$\int_0^\infty \eta e^{\eta s - \eta^2/2} \Gamma(d\eta)$$

$$\leq \int_0^{3s} \eta e^{\eta s - \eta^2/2} \Gamma(d\eta) + \int_{3s}^\infty \eta e^{\eta s - \eta^2/2} \Gamma(d\eta)$$

$$\leq 3s \int_0^\infty e^{\eta s - \eta^2/2} \Gamma(d\eta) + \int_0^\infty \eta e^{-\eta/2} \Gamma(d\eta),$$

since $\eta \geq 3s$ and $s > 1$ implies

$$s\eta - \eta^2/2 \leq -s\eta/2 \leq -\eta/2.$$

Moreover,

$$\int_0^\infty e^{\eta s - \eta^2/2} \Gamma(d\eta) \geq \int_0^\infty \eta e^{-\eta^2/2} \Gamma(d\eta).$$

Therefore,

$$F_\Gamma(s) \leq 3s + a(\Gamma), \qquad (L.29)$$

with

$$a(\Gamma) := \int_0^\infty \eta e^{-\eta/2} \Gamma(d\eta) / \int_0^\infty \eta e^{-\eta^2/2} \Gamma(d\eta).$$

Since a is vaguely continuous,

$$U_0'' := \{\Gamma \in \mathcal{G}_* : a(\Gamma) < 2a(\Gamma_*)\}$$

is open.

By (L.28), (L.29), there exists a constant c such that $\Gamma \in U_0' \cap U_0''$ implies

$$F_\Gamma(s) \leq c(1 + |s|) \qquad \text{for all } s \in \mathbb{R}.$$

□

Lemma L.30. *Let $\mathcal{G}$ denote the family of all p–measures Γ on $\mathbb{B}_+$. Let $\beta \geq 0$ and $\ell \geq 0$.*

The function

$$\Gamma \to \int_0^\infty s^{\beta+\ell} \frac{\left(\int_0^\infty \eta^{\beta+1} \exp[-\eta s]\Gamma(d\eta)\right)^2}{\int_0^\infty \eta^\beta \exp[-\eta s]\Gamma(d\eta)} \, ds \tag{L.31}$$

is continuous on $\mathcal{G}$ with respect to the weak topology if and only if $\ell = 1$.

Proof. (i) Since the weak topology on $\mathcal{G}$ is metrizable, it suffices to consider the convergence of sequences.

Let

$$f(s,\Gamma) := \frac{\left(\int_0^\infty \eta^{\beta+1} \exp[-\eta s]\Gamma(d\eta)\right)^2}{\int_0^\infty \eta^\beta \exp[-\eta s]\Gamma(d\eta)}$$

and

$$g(s,\Gamma) := \int_0^\infty \eta^{\beta+2} \exp[-\eta s]\Gamma(d\eta).$$

By Schwarz's inequality, $f(s,\Gamma) \leq g(s,\Gamma)$. We have to show that $\Gamma_n \Rightarrow \Gamma_0$ implies

$$\int_0^\infty f(s,\Gamma_n) s^{\beta+1} ds \to \int_0^\infty f(s,\Gamma_0) s^{\beta+1} ds. \tag{L.32}$$

Since $\alpha \geq 0$ implies $(\eta \to \eta^\alpha \exp[-\eta s]) \in \mathcal{C}(\mathbb{R}_+)$ for every $s \in \mathbb{R}_+$, the convergence $\Gamma_n \Rightarrow \Gamma_0$ implies $f(s,\Gamma_n) \to f(s,\Gamma_0)$ and $g(s,\Gamma_n) \to g(s,\Gamma_0)$ for every $s \in \mathbb{R}_+$.

Moreover, $\int_0^\infty g(s,\Gamma) s^{\beta+1} ds$ is the same for all $\Gamma \in \mathcal{G}$. Hence (L.32) follows from a suitable version of the Dominated Convergence Theorem (see e.g. Pratt, 1960).

(ii) To see that (L.31) fails to be continuous on $\mathcal{G}$ if $\ell \neq 1$, consider $\Gamma_n := (1-\varepsilon_n)\Gamma_0 + \varepsilon_n \delta_n$, where Γ_0 is a p–measure fulfilling $\int_0^\infty \eta^{1-\ell}\Gamma_0(d\eta) < \infty$, and δ_n is the Dirac–measure assigning probability 1 to $\{n\}$ if $\ell < 1$, and to $\{n^{-1}\}$ if $\ell > 1$. Then one can choose $\varepsilon_n = o(n^0)$ (thus achieving $\Gamma_n \Rightarrow \Gamma_0$) such that $\int_0^\infty f(s,\Gamma_n) s^{\beta+\ell} ds \to \infty$, whereas $\int_0^\infty f(s,\Gamma_0) s^{\beta+\ell} ds < \infty$.

□

The following lemma gives conditions for the consistency of the e.s. defined by (5.29).

Consistency–Lemma L.33. *Let Θ be a Hausdorff space with countable base, A a compact Hausdorff space with countable base. Assume there exists a unique $\alpha_0 \in A$ such that*

$$\frac{P(N_2(\cdot,\vartheta_0,\alpha_0))}{(P(N_1(\cdot,\vartheta_0,\alpha_0)))^2} = \inf_{\alpha\in A} \frac{P(N_2(\cdot,\vartheta_0,\alpha))}{(P(N_1(\cdot,\vartheta_0,\alpha)))^2}\;. \tag{L.34}$$

Assume that the functions

$$(x,\vartheta,\alpha) \to N_2(x,\vartheta,\alpha) \tag{L.35$'$}$$

and

$$(x,\vartheta,\alpha) \to N_1(x,\vartheta,\alpha) \tag{L.35$''$}$$

fulfill condition $L.5((\vartheta_0,\beta),P)$ for all $\beta \in A$.

To avoid technicalities, assume that for all $\beta \in A$

$$P(N_2(\cdot,\vartheta_0,\beta)) > 0 \tag{L.36$'$}$$

and

$$P(N_1(\cdot,\vartheta_0,\beta)) < 0. \tag{L.36$''$}$$

Assume that $\vartheta^{(n)} : X^n \to \Theta$ fulfills

$$\vartheta^{(n)} = \vartheta_0 + o_p(n^0) \quad (P^n). \tag{L.37}$$

Let $A_n \subset A$, $n \in \mathbb{N}$, be a nondecreasing sequence such that $\bigcup_1^\infty A_n$ is dense in A. We consider e.s. $\alpha^{(n)} : X^n \to A_n$, $n \in \mathbb{N}$, with the following property.

$$\varlimsup_{n\to\infty} \Big(n^{-1} \sum_1^n N_2(x_\nu, \vartheta^{(n)}(\underline{x}), \alpha^{(n)}(\underline{x})) / \big(n^{-1} \sum_1^n N_1(x_\nu, \vartheta^{(n)}(\underline{x}), \alpha^{(n)}(\underline{x})) \big)^2$$
$$- \inf_{\alpha \in A_n} n^{-1} \sum_1^n N_2(x_\nu, \vartheta^{(n)}(\underline{x}), \alpha) / \big(n^{-1} \sum_1^n N_1(x_\nu, \vartheta^{(n)}(\underline{x}), \alpha) \big)^2 \Big) = 0. \tag{L.38}$$

The assertion is that any e.s. $\alpha^{(n)}$, $n \in \mathbb{N}$, fulfilling (L.38) converges in probability to the value α_0 fulfilling (L.34).

The intended application is to $N_2(\cdot,\vartheta,\alpha) = N(\cdot,\vartheta,\alpha)^2$ which warrants condition (L.36$'$), and to $N_1(\cdot,\vartheta,\alpha) = N^\bullet(\cdot,\vartheta,\alpha)$. In Examples 1 and 2, the function

$N^\bullet(\cdot,\vartheta,\beta)$ is negative for all $\beta > 0$, which implies (L.36″), a condition which is likely to hold under more general conditions. Because of (5.4) and Remark 5.30, condition (L.36″) follows from $P_{\vartheta_0,\tau_0}(N(\cdot,\vartheta_0,\alpha)L(\cdot,\vartheta_0,\tau_0)) > 0$, $\alpha \in A$, which holds for reasonable choices of $N(\cdot,\vartheta,\alpha)$.

Proof. To simplify our notations we introduce

$$r^{(n)}(\underline{x},\vartheta,\alpha) := n^{-1}\sum_1^n N_2(x_\nu,\vartheta,\alpha) \big/ \Big(n^{-1}\sum_1^n N_1(x_\nu,\vartheta,\alpha)\Big)^2, \tag{L.39}$$

$$r(\vartheta,\alpha) := P(N_2(\cdot,\vartheta,\alpha)) \big/ \big(P(N_1(\cdot,\vartheta,\alpha))\big)^2. \tag{L.40}$$

Since the value α_0 fulfilling (L.34) is assumed to be unique, we have

$$r(\vartheta_0,\beta) > r(\vartheta_0,\alpha_0) \quad \text{for } \beta \neq \alpha_0.$$

Hence there exists $\delta_\beta \in (0,1)$ such that

$$r'_\beta := r(\vartheta_0,\beta)\delta_\beta^3 > r(\vartheta_0,\alpha_0)\delta_\beta^{-3} =: r''_\beta. \tag{L.41}$$

By Lemma L.7 there exist open neighborhoods $U_\beta \ni \vartheta_0$, $V_\beta \ni \alpha_0$, $W_\beta \ni \beta$ such that (recall (L.36′))

$$P(\underline{N}_2(\cdot,U_\beta,W_\beta)) > \delta_\beta P(N_2(\cdot,\vartheta_0,\beta)), \tag{L.42′}$$

$$P(\overline{N}_2(\cdot,U_\beta,V_\beta)) < \delta_\beta^{-1} P(N_2(\cdot,\vartheta_0,\alpha_0)) \tag{L.42″}$$

and (recall (L.36″))

$$0 > P(\underline{N}_1(\cdot,U_\beta,W_\beta)) > \delta_\beta^{-1} P(N_1(\cdot,\vartheta_0,\beta)), \tag{L.43′}$$

$$P(\overline{N}_1(\cdot,U_\beta,V_\beta)) < \delta_\beta P(N_1(\cdot,\vartheta_0,\alpha_0)) < 0. \tag{L.43″}$$

By the strong law of large numbers there exists a $P^{\mathbb{N}}$–null set $N_\beta \subset X^{\mathbb{N}}$ such that the relations (L.42) and (L.43) hold for $\underline{x} \notin N_\beta$ with $P(F)$ on the left–hand side replaced by $\lim_{n\to\infty} n^{-1}\sum_1^n F(x_\nu)$. Hence we obtain for $\underline{x} \notin N_\beta$ (see (L.41))

$$\lim_{n\to\infty} \frac{n^{-1}\sum_1^n \underline{N}_2(x_\nu,U_\beta,W_\beta)}{\big(n^{-1}\sum_1^n \underline{N}_1(x_\nu,U_\beta,W_\beta)\big)^2} > r'_\beta, \tag{L.44′}$$

$$\lim_{n\to\infty} \frac{n^{-1}\sum_1^n \overline{N}_2(x_\nu,U_\beta,V_\beta)}{\big(n^{-1}\sum_1^n \overline{N}_1(x_\nu,U_\beta,V_\beta)\big)^2} < r''_\beta, \tag{L.44″}$$

hence also (see (L.39))

$$\varliminf_{n\to\infty} r^{(n)}(\underline{x}, U_\beta, W_\beta) > r'_\beta \qquad (L.45')$$

and

$$\varlimsup_{n\to\infty} \bar{r}^{(n)}(\underline{x}, U_\beta, V_\beta) < r''_\beta. \qquad (L.45'')$$

Let V_0 be an open neighborhood of α_0. Since A is compact, so is $A \cap V_0^c$. Since $\{W_\beta : \beta \in A \cap V_0^c\}$ covers the compact set $A \cap V_0^c$, it contains a finite subcover, say $W_{\beta_1}, \ldots, W_{\beta_r}$. Let $N_* := \bigcup_{i=1}^{r} N_{\beta_i}$.

Since $\vartheta^{(n)}$, $n \in \mathbb{N}$, converges to α_0 in probability, for any subsequence $\mathbb{N}_0 \subset \mathbb{N}$ there exists a $P^\mathbb{N}$-null set N_0 and a subsequence $\mathbb{N}_1 \subset \mathbb{N}_0$ such that $\lim_{n\in\mathbb{N}_1} \vartheta^{(n)}(\underline{x}) = \vartheta_0$ for $\underline{x} \notin N_0$. We shall show that $\alpha^{(n)}(\underline{x})$, $n \in \mathbb{N}_1$, is eventually in V_0 for any $\underline{x} \notin N_0 \cup N_*$. Let $\underline{x} \notin N_0 \cup N_*$ be fixed. If $\alpha^{(n)}(\underline{x}) \in V_0^c$ infinitely often, then there exists $i_0 \in \{1, \ldots, r\}$ and an infinite subsequence $\mathbb{N}_2 \subset \mathbb{N}_1$ (both depending on $\underline{x}$) such that $\alpha^{(n)}(\underline{x}) \in W_{\beta_{i_0}}$ for $n \in \mathbb{N}_2$.

In this case,

$$\varliminf_{n\in\mathbb{N}_2} r^{(n)}\big(\underline{x}, \vartheta^{(n)}(\underline{x}), \alpha^{(n)}(\underline{x})\big) \geq \varliminf_{n\in\mathbb{N}} r^{(n)}(\underline{x}, U_{\beta_{i_0}}, W_{\beta_{i_0}}) > r'_{\beta_{i_0}}. \qquad (L.46)$$

Moreover,

$$\inf_{\alpha\in A_n} r^{(n)}\big(\underline{x}, \vartheta^{(n)}(\underline{x}), \alpha\big) \leq \bar{r}^{(n)}(\underline{x}, U_{\beta_{i_0}}, V_{\beta_{i_0}}),$$

if $n \in \mathbb{N}_2$ is large enough so that $\vartheta^{(n)}(\underline{x}) \in U_{\beta_{i_0}}$ and $A_n \cap V_{\beta_{i_0}} \neq \emptyset$.

Hence

$$\varlimsup_{n\in\mathbb{N}_2} \inf_{\alpha\in A_n} r^{(n)}\big(\underline{x}, \vartheta^{(n)}(\underline{x}), \alpha\big) \leq \varlimsup_{n\in\mathbb{N}} \bar{r}^{(n)}(\underline{x}, U_{\beta_{i_0}}, V_{\beta_{i_0}}) < r''_{\beta_{i_0}}. \qquad (L.47)$$

From (L.46) and (L.47)

$$\varliminf_{n\in\mathbb{N}_2} \big[r^{(n)}\big(\underline{x}, \vartheta^{(n)}(\underline{x}), \alpha^{(n)}(\underline{x})\big) - \inf_{\alpha\in A_n} r^{(n)}\big(\underline{x}, \vartheta^{(n)}(\underline{x}), \alpha\big)\big] > r'_{\beta_{i_0}} - r''_{\beta_{i_0}} > 0,$$

in contradiction to (L.38).

This proves that $\alpha^{(n)}(\underline{x})$, $n \in \mathbb{N}_1$, is eventually in V_0 for all $\underline{x} \notin N_0 \cup N_*$.

The null set N_* still depends on the particular neighborhood V_0. Since α_0 admits a countable neighborhood system, this implies that $\alpha^{(n)}(\underline{x})$, $n \in \mathbb{N}_1$, converges to α_0 for $P^\mathbb{N}$-a.a. $\underline{x} \in X^\mathbb{N}$.

Since the subsequence $\mathbb{N}_0$ was arbitrary, this implies that $\alpha^{(n)}$, $n \in \mathbb{N}$, converges to α_0 in probability (P^n).

□

Acknowledgment

The author wishes to thank R. Hamböker and L. Schröder who checked the technical details and suggested various improvements, and W. Wefelmeyer whose advice contributed to make the paper more readable. A. Schick suggested, among other things, an improvement of Lemma 3.18.

The numerical computations were started by W. Krimmel and continued by R. Hamböker, using the CYBER 76M at the *Rechenzentrum der Universität zu Köln*.

Finally, the author is grateful to the Department of Statistics and Probability at Michigan State University, East Lansing, in particular Professor V. Fabian, who enabled him to give a series of lectures on this subject in December 1988.

The TeX version was done by E. Lorenz.

References

Amari, S. (1987a). Differential geometrical theory of statistics. In: *Differential Geometry in Statistical Inference* (S. Amari et al., ed.), 19–94, Lecture Notes–Monograph Series **10**, Inst. Math. Statist.

Amari, S. (1987b). Dual connections on the Hilbert bundles of statistical models. In: *Geometrization of Statistical Theory* (C.T.J. Dodson, ed.), 123–151, Proceedings of the GST Workshop, Lancaster 1987.

Amari, S. and Kumon, M. (1988). Estimation in the presence of infinitely many nuisance parameters — geometry of estimating functions. *Ann. Statist.* **16** 1044–1068.

Andersen, E.B. (1970). Asymptotic properties of conditional maximum likelihood estimators. *J. Roy. Statist. Soc. Ser. B* **32** 283–301.

Bauer, H. (1981). *Probability Theory and Elements of Measure Theory*. Acadmic Press, New York.

Begun, J.M., Hall, W.J., Huang, W.-M. and Wellner, J.A. (1983). Information and asymptotic efficiency in parametric–nonparametric models. *Ann. Statist.* **11** 432–452.

Beran, R. (1978). An efficient and robust adaptive estimator of location. *Ann. Statist.* **6** 292–313.

Bickel, P.J. (1981). Quelques aspects de la statistique robuste. In: *Ecole d'Eté de Probabilités de Saint–Flour IX–1979* (P.L.Hennequin, ed.), 1–72, Lecture Notes in Mathematics 876, Springer–Verlag, Berlin.

Bickel, P.J. (1982). On adaptive estimation. *Ann. Statist.* **10** 647–671.

Bickel, P.J. and Ritov, Y. (1987). Efficient estimation in the errors in variables model. *Ann. Statist.* **15** 513–540.

Bickel, P.J. and Ritov, Y. (1989). Achieving information bounds in non and semiparametric models. To appear: *Ann. Statist.*

Bickel, P.J.,Klaassen, C.A.J., Ritov, Y. and Wellner, J.A. (199?). *Efficient and Adaptive Inference in Semiparametric Models*. Johns Hopkins University Press, Baltimore. To appear.

Blum, J.R. and Susarla, V. (1977). Estimation of a mixing distribution function. *Ann. Probab.* **5** 200–209.

Borges, R. and Pfanzagl, J. (1965). One–parameter exponential families generated by transformation groups. *Ann. Math. Statist.* **36** 261–271.

Choi, K. (1969). Estimators for the parameters of a finite mixture of distributions. *Ann. Inst. Statist. Math.* **21** 107–116.

Cox, D.R. and Hinkley, D.V. (1974). *Theoretical Statistics.* Chapman and Hall, London.

Deeley, J.J. and Kruse, R.L. (1968). Construction of sequences estimating the mixing distribution. *Ann. Math. Statist.* **39** 286–288.

Godambe, V.P. (1976). Conditional likelihood and unconditional optimum estimating equations. *Biometrika* **63** 277–284.

Godambe, V.P. and Thompson, M.E. (1974). Estimating equations in the presence of a nuisance parameter. *Ann. Statist.* **2** 568–571.

Gong, G. and Samaniego, F.J. (1981). Pseudo maximum likelihood estimation: theory and applications. *Ann. Statist.* **9** 861–869.

Hájek, J. (1962). Asymptotically most powerful rank–order tests. *Ann. Math. Statist.* **33** 1124–1147.

Hall, P. (1981). On the nonparametric estimation of mixture proportions. *J. Roy. Statist. Soc. Ser. B* **43** 147–156.

Kalbfleisch, J. and Sprott, D.A. (1970). Application of likelihood methods to models involving large numbers of parameters (with discussion). *J. Roy. Statist. Soc. Ser. B* **32** 175–208.

Klaassen, C.A.J. (1987). Consistent estimation of the influence function of locally asymptotically linear estimators. *Ann. Statist.* **15** 1548–1562.

Koshevnik, Yu. A. and Levit, B. Ya. (1976). On a nonparametric analogue of the information matrix. *Theor. Probab. Appl.* **21** 738–753.

Kumon, M. and Amari, S. (1984). Estimation of a structural parameter in the presence of a large number of nuisance parameters. *Biometrika* **71** 445–459.

LeCam, L. (1960). Locally asymptotically normal families of distributions. *Univ. Calif. Publ. Statist.* **3**, 37–98.

Lehmann, E.L. (1986). *Testing Statistical Hypotheses.* 2nd ed., Wiley, New York.

Levit, B. Ya. (1974). On optimality of some statistical estimates. In: *Proceedings of the Prague Symposium on Asymptotic Statistics*, Vol. 2 (J. Hájek, ed.), 215–238, Charles University, Prague.

Levit, B. Ya. (1975). On the efficiency of a class of non–parametric estimates. *Theor. Probab. Appl.* **20** 723–740.

Liang, K.-Y. (1983). On information and ancillarity in the presence of a nuisance parameter. *Biometrika* **70** 604–612.

Lindsay, B. (1982). Conditional score functions: some optimality results. *Biometrika* **69** 503–512.

Lindsay, B. (1983). Efficiency of the conditional score in a mixture setting. *Ann. Statist.* **11** 486–497.

Lindsay, B. (1985). Using empirical partially Bayes inference for increased efficiency. *Ann. Statist.* **13** 914–931.

Little, R.J.A. and Rubin, D.B. (1987). *Statistical Analysis with Missing Data.* Wiley, New York.

Morton, R. (1981). Efficiency of estimating equations and the use of pivots. *Biometrika* **68** 227–233.

Neyman, J. and Scott, E.L. (1948). Consistent estimates based on partially consistent observations. *Econometrica* **16** 1–32.

Parthasarathy, K. (1967). *Probability Measures on Metric Spaces.* Academic Press, New York.

Pfanzagl, J. (1987). Bounds for the asymptotic efficiency of estimators based on functional contractions; applications to the problem of estimation in the presence of random nuisance parameters. In: *Proceedings of the 1st World Congress of the Bernoulli Society* (Yu. Prohorov, V.V. Sazonov, eds.), Vol. 2. *Probability Theory and Mathematical Statistics*, 237–248, VNU Science Press, Utrecht.

Pfanzagl, J. (1988). Consistency of maximum likelihood estimators for certain nonparametric families, in particular: mixtures. *J. Statist. Plann. Inference* **19** 137–158.

Pfanzagl, J. (1989). Large deviation probabilities for certain nonparametric maximum likelihood estimators. Preprints in Statistics 118, Math. Institute, University of Cologne. To appear: *Ann. Statist.* 1990.

Pfanzagl, J. and Wefelmeyer, W. (1982). *Contributions to a General Asymptotic Statistical Theory.* Lecture Notes in Statistics, Vol. 13, Springer–Verlag, New York.

Pfanzagl, J. and Wefelmeyer, W. (1985). *Asymptotic Expansions for General Statistical Models.* Lecture Notes in Statistics, Vol. 31, Springer–Verlag, Berlin.

Prakasa Rao, B.L.S. (1987). *Asymptotic Theory of Statistical Inference.* Wiley, New York.

Pratt, J.W. (1960). On interchanging limits and integrals. *Ann. Math. Statist.* **31** 74–77.

Rieder, H. (1983). Robust estimation of one real parameter when nuisance parameters are present. In: Transactions of the Ninth Prague Conference on Information Theory, Statistical Decision Functions, Random Processes, Vol. A (Prague 1982), 77–89, Reidel, Dordrecht–Boston, Mass.

Robbins, H. (1963). The empirical Bayes approach to testing statistical hypotheses. *Rev. Inst. Internat. Statist.* **31** 195–208.

Robbins, H. (1964). The empirical Bayes approach to statistical decision problems. *Ann. Math. Statist.* **35** 1–20.

Schick, A. (1986). On asymptotically efficient estimation in semiparametric models. *Ann. Statist.* **14** 1139–1151.

Schick, A. (1987). A note on the construction of asymptotically linear estimators. *J. Statist. Plann. Inference* **16** 89–105. Correction **22** (1989) 269–270.

Singh, R.S. (1979). Empirical Bayes estimation in Lebesgue–exponential families with rates near the best possible rate. *Ann. Statist.* **7** 890–902.

Sprent, P. (1969). *Models in Regression and Related Topics.* Methuen, London.

Stone, C. (1975). Adaptive maximum likelihood estimators of a location parameter. *Ann. Statist.* **3** 267–284.

Titterington, D.M. (1989). Some recent research in the analysis of mixture distributions. To appear: *Statistics.*

van der Vaart, A. (1988). *Statistical Estimation in Large Parameter Spaces.* CWI Tract 44, Stichting Mathematisch Centrum, Amsterdam.

Notation index

as.	asymptotic
$a(\cdot, \vartheta)$	40 (7.6′)
$b(\cdot, \vartheta)$	40 (7.6″)
$B(s), \hat{B}(s)$	55, 56 (9.12)
$I\!B, I\!B^k$	Borel algebra over $I\!R$ or $I\!R^k$
$I\!B_+$	$I\!B \cap I\!R_+$
$\mathcal{C}(H)$	class of all bounded and continuous functions on H
$\mathcal{C}_0(H)$	class of all continuous functions on H vanishing at infinity
$C(s, \vartheta), \hat{C}(s, \vartheta)$	55 (9.10)
$\vartheta_*^{(n)}$	discretized estimator 43 (7.14)
e.s.	estimator sequence
$f^\bullet(x, \vartheta)$	$\frac{\partial}{\partial \vartheta} f(x, \vartheta)$
$f'(x, \vartheta)$	$\frac{\partial}{\partial x} f(x, \vartheta)$
$\overline{f}(x, B), \underline{f}(x, B)$	88
$f^{(n)} \to f \ (P^n)$	stochastic convergence
$\mathcal{G}$	a class of p-measures on $(H, \mathcal{C})$
$\mathcal{G}_*$	class of all sub-probability measures on $(H, \mathcal{C})$
$\Gamma_{\alpha, \beta}$	61 (9.28)
$H_0(s, \Gamma)$	53 (9.4)
$H_2(s, \Gamma)$	59 (9.24)
$H'(s, \alpha)$	$\frac{\partial}{\partial s} H(s, \alpha)$
$h^{(i)}(\alpha_0, \ldots, \alpha_m)$	14
$\kappa^+(\cdot, P)$	gradient 3
$\kappa^*(\cdot, P)$	canonical gradient 3

$k(\cdot, P)$	7
$K(\cdot, P)$	influence function (2.2)
$\ell^{(i)}(\cdot, \vartheta, \tau)$	20
$L_{ij}(\vartheta, \tau)$	20
$L(\cdot, \vartheta, \tau)$	18
$L_0(\cdot, \vartheta, \tau)$	40 (7.7)
$\mathcal{L}_*(P)$	2
$\mathcal{L}_S(P_{\vartheta,\tau})$	39
$\Lambda(\vartheta, \tau)$	21
λ, λ^k	Lebesgue measure on $I\!B$ or $I\!B^k$, respectively
m.l.	maximum likelihood
$o_p(n^{-\alpha}), O_p(n^{-\alpha})$	generic sequence $R_n : X^n \to I\!R$ such that $n^\alpha R_n$ converges stochastically to zero or is stochastically bounded, respectively
p–measure	probability measure
$P * T$	distribution of T under P
$P_n \Rightarrow P$	weak convergence
$P_t \to P$	2
$P_{\vartheta,\cdot}^{S(\cdot,\vartheta)}$	35
$R_{\vartheta,\tau}$	38
$I\!R_+$	$(0, \infty)$
$T(P, \mathcal{P})$	tangent space 3. Also $T(P)$, if $\mathcal{P}$ is understood.
$T_0(P_{\vartheta,\tau})$	level space 17
$T_0^\perp(P_{\vartheta,\tau})$	orthogonal complement of $T_0(P_{\vartheta,\tau})$ in the tangent space

I	Pfanzagl, J. and Wefelmeyer, W. (1982)
II	Pfanzagl, J. and Wefelmeyer, W. (1985)

Subject index

adaptivity	18
as. variance bound	$\underline{4}$, 7, 18, 37, 41, 55, 69, 70, 71, 77, 78, 80, 81, 84
asymptotically linear	$\underline{4}$, 5, 6, 7, 8, 12
completeness	39, 48
Convolution Theorem	4, 6
differentiable functional	$\underline{3}$, 4, 14
discretized estimator	19, 43, 47, 50, 57, 96
distribution	
Gamma —	55, 61, 62, 67, 69, 72, 73, 74, 75, 78, 80
mixing —	49, 61
normal —	4, 5, 55, 75, 82, 84, 85, 86, 87
estimating equation	22, 35, 36, 37, 66, 84, 85
estimating function	35, 36, 37, 70, 71, 79
full family	39, 40
gradient	$\underline{3}$, 4, 5, 6, 7, 10, 14, 15, 17, 18, 23, 26, 36, 37
canonical —	$\underline{3}$, 7, 10, 14, 17, 18, 19, 24, 41, 46
improvement procedure	8, 21, 25, 46, 50, 51, 56, 57, 59, 65, 66; Examples 1–3, 67–87
influence function	$\underline{4}$, 6, 7, 8, 12, 18, 19
kernel estimator	46, 47, 49, 71
level space	$\underline{17}$, 21, 22, 37, 39, 48
maximum likelihood estimator	34, 41, 42, 49, 50, 51, 52, 59, 60, 64
mixture models	37, 39, 42, 46, 47, 48, 49, 53; Examples 1–3, 67–87
parametric families	20, 22, 61
permutation invariance	7, 8, 9, 12, 26, 27, 28, 29, 31, 93, 94
semiparametric models	9, 11, 17, 22
simulation experiments	47, 49, 63, 64, 66; Examples 1–3, 67–87
splitting	10, 11, 12, 19, 42
sufficient statistic	35, 37, 38, 39, 45, 48, 51, 53, 55, 61
tangent space	$\underline{3}$, 4, 17, 21, 39